SpringerBriefs in Bioengineering

SpringerBriefs present concise summaries of cutting-edge research and practical applications across a wide spectrum of fields. Featuring compact volumes of 50 to 125 pages, the series covers a range of content from professional to academic. Typical topics might include: A timely report of state-of-the art analytical techniques, a bridge between new research results, as published in journal articles, and a contextual literature review, a snapshot of a hot or emerging topic, an in-depth case study, a presentation of core concepts that students must understand in order to make independent contributions.

Krish W. Ramadurai • Abhirup Banerjee

Machine Learning-Driven Rational Design in Nanomedicine

Advances in Computational Drug Delivery and In Silico Screening

Krish W. Ramadurai (iD)
Department of Engineering Science
Institute of Biomedical Engineering
University of Oxford
Oxford, UK

Abhirup Banerjee
Department of Engineering Science
Institute of Biomedical Engineering
University of Oxford
Oxford, UK

ISSN 2193-097X   ISSN 2193-0988   (electronic)
SpringerBriefs in Bioengineering
ISBN 978-3-032-04011-4   ISBN 978-3-032-04012-1   (eBook)
https://doi.org/10.1007/978-3-032-04012-1

This Springer imprint is published by the registered company Springer Nature Switzerland AG
The registered company address is: Gewerbestrasse 11, 6330 Cham, Switzerland

If disposing of this product, please recycle the paper.

Competing Interests The authors have no competing interests to declare that are relevant to the content of this manuscript.

About the Book

One of the most critical bottlenecks in developing the next generation of nanomedicines, such as mRNA oncological vaccines, is the availability of suitable nanocarriers with adequate transfection efficiency for sufficient payload delivery. It is challenging to find a safer and more efficient nanocarrier because a suitable nanocarrier requires identifying materials that effectively combine with the drugs, avoid rejection by human cells, and have low toxicity. Traditional laboratory research and development for nanocarriers have relied upon manual workflows that are slow and inefficient. With machine learning (ML) enhanced virtual screening, this process can finally transition from a laborious, trial-and-error process to a highly automated and precise design process. This research explores the optimization and application of advanced ML models integrated with chemoinformatic data computation techniques, including molecular fingerprinting, to enable the high-throughput in silico screening of lipid nanoparticles (LNPs). Molecular fingerprinting transforms molecular structures into numerical representations that capture essential chemical and structural data, generating detailed molecular profiles that facilitate the systematic exploration of how different molecular features influence the biological activity and efficacy of LNPs. Integrating these fingerprints into ML models can facilitate the precise prediction of LNP properties, such as transfection efficiency and biocompatibility, for use in nanomedicines, such as nano-oncological vaccines. This integration enhances the ability to identify optimal LNP formulations, potentially reducing the time and cost associated with traditional experimental methods. This approach can ultimately provide researchers with a tool to design highly effective therapies, thereby streamlining the development of next-generation nanomedicines.

Contents

Abbreviations

AUC	Area under the curve
caTLR4	Constitutively active toll-like receptor 4
CD40L	Cluster of differentiation 40 ligand
DLin-MC3-DMA	Dilinoleylmethyltrimethylaminopropane
DMG-PEG2000	Dimyristoyl glycerol polyethylene glycol 2000
DODMA	1,2-Dioleoyloxy-3-(trimethylammonium)propane
DOPE	Dioleoylphosphatidyl ethanolamine
DOTMA	Dioleoyltrimethylammonium propane
DSPC	Distearoylphosphatidylcholine
ECFPs	Extended-connectivity fingerprints
ENN	Edited nearest neighbors
EDA	Exploratory data analysis
IL-12	Interleukin 12
IL-2	Interleukin 2
LNP	Lipid nanoparticle
MAGE	Melanoma antigen gene
MoAs	Mechanisms of action
mRNA	Messenger ribonucleic acid
NPs	Nanoparticles
OX40L	OX40 ligand
PEG	Polyethylene glycol
PSA	Polar surface area
RFC	Random forest classifier
ROC	Receiver operating characteristic
RWE	Real-world evidence
SARS-CoV-2	Severe acute respiratory syndrome coronavirus 2
SMILES	Simplified molecular input line entry system
SMOTE	Synthetic minority over-sampling technique
SSL	Semi-supervised learning

TAAs	Tumor-associated antigens
TE	Transfection efficiency
TSAs	Tumor-specific antigens
VAE	Variational autoencoder

Chapter 1
An Introduction to mRNA Vaccines in Cancer Nanomedicine

The current interventional treatment landscape for oncology largely relies upon conventional clinical treatments for tumors, including surgery, chemotherapy, immunotherapy, or combinatorial therapeutic protocols. Despite advancements in these treatment options, malignant cancers remain a leading cause of mortality globally (Sung et al., 2021; Yu et al., 2015). Furthermore, the development of oncology drugs continues to be plagued by an attrition rate greater than 95%, with poor clinical translatability in patients (Jentzsch et al., 2023). Recent advances have focused on an endogenous approach via tumor immunotherapies that activate the patient's immune system and antitumor immunity, leading to a tumor-suppressive microenvironment and improving survival outcomes (Zang et al., 2017). One specific area of interest that embodies this approach is nano-oncological vaccines, which offer a unique approach to tumor treatment and immunotherapy. These vaccines utilize tumor-associated antigens (TAAs) or tumor-specific antigens (TSAs) that can eliminate malignant tumor cells by inducing the expression of these antigens and achieving durable treatment responses through the development of immune memory (Zang et al., 2017). Cancer vaccines could provide a highly efficacious and durable treatment response, leading to tumor regression and elimination, with fewer adverse side effects compared to other immunotherapies and chemotherapies (Zhao et al., 2019).

The emergence of mRNA oncology vaccines represents a potentially revolutionary shift in cancer treatment, in which these vaccines leverage messenger RNA (mRNA) to instruct cells to produce proteins foreign to the body, thereby eliciting an immune response capable of targeting and eliminating cancer cells (Deng et al., 2022). mRNA oncology vaccines can encode and express TSA and TAAs, stimulating both humoral and cellular immunity that can lead to tumor regression in patients (Deng et al., 2022). Furthermore, unlike DNA vaccines, mRNA does not integrate into the host genome and cause mutagenesis (Park et al., 2021). However, several challenges must be overcome to research and develop these breakthrough mRNA

© The Author(s), under exclusive license to Springer Nature
Switzerland AG 2026
K. W. Ramadurai, A. Banerjee, *Machine Learning-Driven Rational Design in
Nanomedicine*, SpringerBriefs in Bioengineering,
https://doi.org/10.1007/978-3-032-04012-1_1

nano-oncological vaccines effectively. The first issue relates to delivery mechanics, in which the mRNA needs to be directed and integrated into the desired cells without degradation. To address this, mRNA can be encapsulated into lipid nanoparticles (LNPs) or liposomes to protect it (Ramachandran et al., 2022). The design of LNP as a vector for delivering mRNA can overcome several disadvantages of mRNA therapy. Specifically, mRNA has difficulty crossing the cellular membrane as a negatively charged macromolecule, has an intracellular half-life of less than 8 hours, and can also be trapped in the endosome of cells, preventing translation (Zhou et al., 2023). Compared to naked mRNA, LNPs can stabilize delivery and alter the pharmacokinetic (PK) properties of drugs to prevent in vivo degradation and enhance tumor uptake (Moreno-Rocha et al., 2012; Zong et al., 2023). Liposomes are a primary nanocarrier for mRNA vaccines, but they can still exhibit cytotoxic properties in the human body (Chen et al., 2023). Liposomes comprise of phospholipid layers that encapsulate an mRNA payload and typically display high biocompatibility and versatility in encapsulating both hydrophilic and hydrophobic substances (Eloy et al., 2014). When it comes to manufacturing and mRNA encapsulation capacity, liposomes are often challenging to manufacture at scale and typically have diminished encapsulation capacity and transfection efficiency relative to LNPs (Tenchov et al., 2021). Given this, most pharmaceutical developers have focused on producing mRNA vaccines with LNP delivery technology.

While the concept of oncology vaccines has been promising for many years, their translation into effective clinical therapies remains poor due to the diversity of tumor antigens, low endogenous host immune response, and insufficient therapeutic delivery mechanics (Tay et al., 2021). When it comes to enhancing the delivery mechanisms of these therapeutics, a significant problem remains in predicting and screening the transfection efficiency (TE) of nanocarriers. TE refers to the ability of the mRNA within the LNPs to enter the cells and facilitate protein production (Huang et al., 2022; Moayedpour et al., 2024; Patel et al., 2023). High TE ensures that a sufficient payload, such as mRNA, is delivered to the target cells and ensures that antigen-presenting cells efficiently take up the mRNA-encoding tumor antigens and activate the patient's immune system (Gómez-Aguado et al., 2020). This enhances the overall effectiveness of the therapy by ensuring that therapeutic proteins or antigens are produced at levels sufficient to stimulate a potent immune response against tumor cells. The ability to screen LNPs with higher TE in silico means less time is spent on formulations that likely would not translate well in biological contexts, saving time and experimental resources for laboratory scientists and researchers.

Given the critical role of TE in mRNA oncology vaccines and nanomedicines, this research focuses on optimizing the in silico prediction of LNP TE using machine learning (ML)- enhanced virtual screening combined with chemoinformatic techniques, including molecular fingerprinting, to address this critical bottleneck. Traditional modalities of laboratory research and development for nanocarriers have relied upon slow, inefficient workflows, in which ML has opened the door to the rapid design and identification of safer, more efficient nanocarriers for the improved efficacy of nano-oncological therapies (Ding et al., 2023; Öztürk et al.,

2018). LNP design is critical for effective mRNA vaccine delivery, as LNPs must be stable and biocompatible with the proper mRNA encapsulation and TE profile. This research hypothesizes that integrating advanced ML virtual screening techniques with structure-based chemoinformatic analyses, including molecular fingerprinting, could enhance the prediction accuracy and in silico screening of LNP transfection efficiency for mRNA oncology vaccines, accelerating the design and development of next-generation nanomedicines. In tackling this research problem, it is essential to understand the distinct molecular and biological dynamics required for successful tumor regression using these vaccines, as well as the unique and distinct classifications of the mechanisms of action (MoAs) of these novel interventional treatments.

1.1   mRNA Vaccine Mechanism of Action Classifications

The unique MoA of mRNA cancer vaccines offers significant advantages over conventional cancer therapies, such as surgery and chemotherapy. One primary advantage is their potent immunogenicity, which elicits robust cell-mediated immune responses (Deng et al., 2022). For metastatic tumors, which are not easily treated by surgery, mRNA cancer vaccines are an effective alternative as they activate systemic immune responses within patients. While the stimulation of this host immune response is important, another critical element is the durability of the response that can result in permanent tumor regression. mRNA cancer vaccines can establish and maintain long-term immunological memory, potentially reducing the risk of tumor recurrence. mRNA cancer vaccines utilize several distinct MoAs, specifically targeting TSAs, TAAs, or encoded immunostimulatory factors (Fig. 1.1) (Deng et al., 2022). mRNA cancer vaccines encoding TSAs are highly immunogenic, effectively stimulating strong T-cell responses by targeting mutations unique to the tumor (Hu et al., 2018; Zong et al., 2023). The mRNA sequences are delivered to immune cells via LNPs, enabling them to identify and destroy cancer cells. TAAs comprise tumor- or germline-expressed cancer genes and lineage-specific differentiation markers, which are excellent targets for vaccine development (Tay et al., 2021). Cancer mRNA vaccines encode TAAs, such as overexpressed tumor antigens in cancer cells, in which mRNA is translated into an antigen presented in the MHC II complex to stimulate T-cell responses against antigen-positive cells (Fig. 1.3a) (He et al., 2022). Cancer mRNA vaccines are especially effective for cancers with high expression of TAA, such as lung cancer and melanoma (He et al., 2022). mRNA vaccines encoding immune-stimulatory factors, such as cytokines interleukin 2 (IL-2) and IL-12, and co-stimulatory ligands, such as OX40 Ligand (OX40L), promote immune system response against tumor cells (Fig. 1.2) (Beck et al., 2021). These vaccines can also serve as adjuvants for TAA-encoding mRNA vaccines, enhancing their immune activity and antitumor effects. For example, ECI-006 is a combination mRNA cancer vaccine comprising TriMix that contains mRNA-encoding dendritic cell activation molecules such as Cluster of Differentiation 40 Ligand (CD40L) and

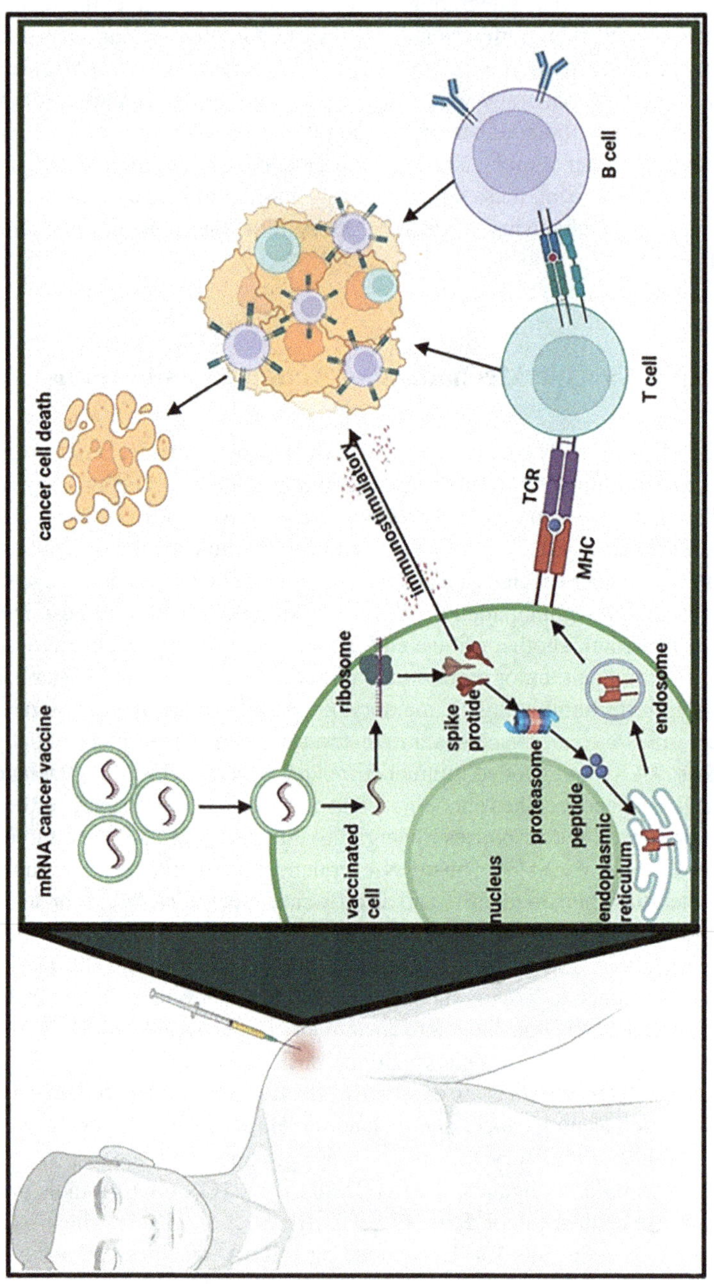

Fig. 1.1 mRNA vaccine MoA for oncological applications (Deng et al., 2022)

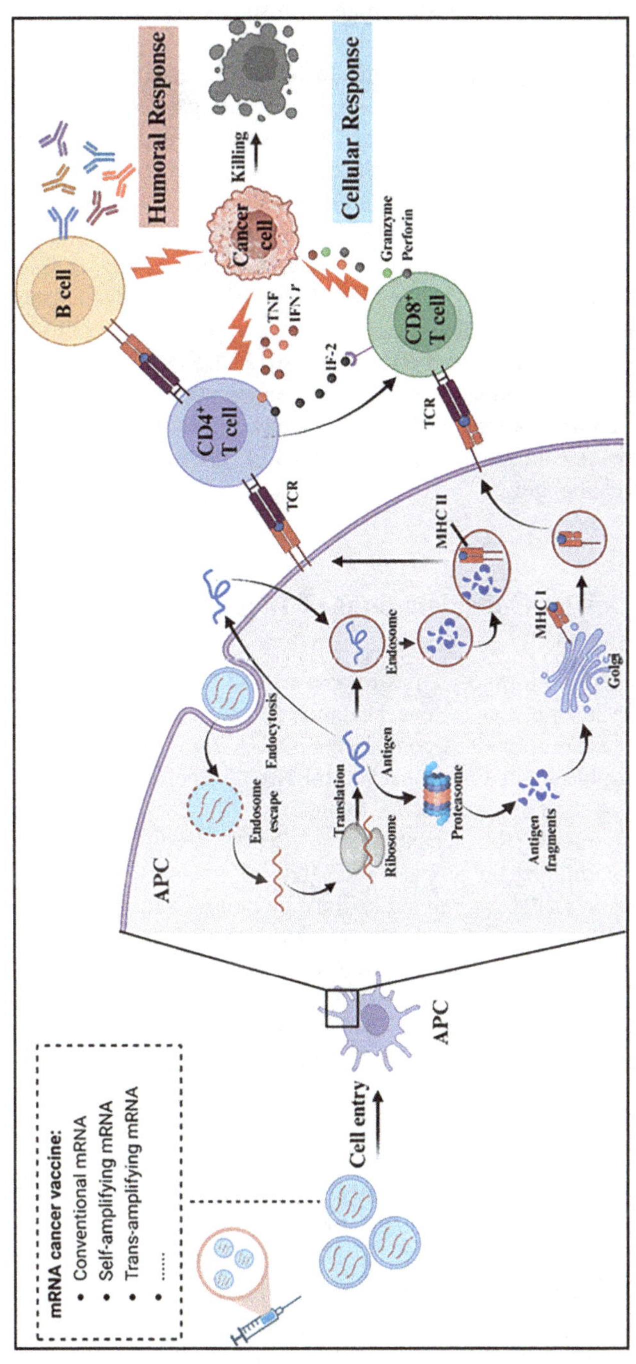

Fig. 1.2   Mechanism of mRNA cancer vaccines in inducing anti-tumor immunity (Fu et al., 2025)

mRNA encoding TAAs such as tyrosinase and Melanoma Antigen Gene (MAGE A3), which show promise in boosting immune responses against cancer cells (Wang et al., 2023).

Current cancer treatment studies indicate the diverse utility of RNA-based therapies across several oncological pathologies, including glioblastoma, melanoma, colorectal cancer, and hepatocellular carcinoma (Meulewaeter et al., 2023). Unlike DNA delivery, mRNA delivery specifically aims to positively regulate protein expression without the risk of insertional mutations, providing consistent protein expression (Gómez-Aguado et al., 2020; Vlahos et al., 2022). Additionally, mRNA drugs are easier to synthesize in vitro, and mRNA displays a higher TE than DNA, particularly in immune cells (Ho et al., 2021). Despite these advantages, the use of naked RNA in vitro faces limitations, including poor chemical stability, short half-life, and degradation by nucleases (Guo et al., 2012). LNPs can stabilize and encapsulate RNA molecules, protecting them from enzymatic degradation and immune system clearance while allowing for targeted delivery (Fang et al., 2022). However, ensuring the overall translation, treatment durability, and efficacy of mRNA vaccines remains challenging.

1.2 LNP Design Considerations of LNPs for mRNA Delivery

While lipid-based platforms, such as lipoplexes and LNPs, are readily utilized for mRNA delivery, their respective encapsulation and molecular interaction properties differ. Lipoplexes are formed by complexing mRNA with cationic liposomes, while LNPs use ionizable lipids to encapsulate mRNA (Schlich et al., 2021). LNPs are preferred due to their higher mRNA expression efficiency and reduced toxicity compared to lipoplexes (Ramachandran et al., 2022). The lipid carrier composition is critical in determining the efficacy and safety of a nano-oncological vaccine, as ionizable lipids in LNPs can reduce toxicity by neutralizing at physiological pH, unlike cationic lipids used in lipoplexes, which can cause inflammation and necrosis (Schlich et al., 2021). Thus, when it comes to the functional design of mRNA-LNP cancer vaccines, a balance must be struck between determining the sufficient ability to trigger immune activation and the desired antitumoral effect in patients (Meulewaeter et al., 2023). The choice of adjuvant, lipid carriers, mRNA modifications, and passive immune targeting must all be considered when creating a viable and efficacious cancer vaccine (Fig. 1.3) (Liu et al., 2023).

LNPs are efficient delivery systems for various clinical pathologies in infectious disease, oncology, and neurodegenerative diseases (Adams et al., 2018; Polack et al., 2020; Pozzi & Caracciolo, 2023). Encapsulation into a lipid shell involves a charge interaction between mRNA's negatively charged phosphate backbone and cationic lipids' positively charged amino headgroups (Zhu & Mahato, 2010). Cationic lipids with a single amino headgroup are positively charged, for example, Dioleoyltrimethylammonium Propane (DOTMA), and the cationic lipids carrying two or more substituents in amino headgroup are referred to as ionizable lipids, such as 1,2-Dioleoyloxy-3-(trimethylammonium)propane (DODMA) with neutral

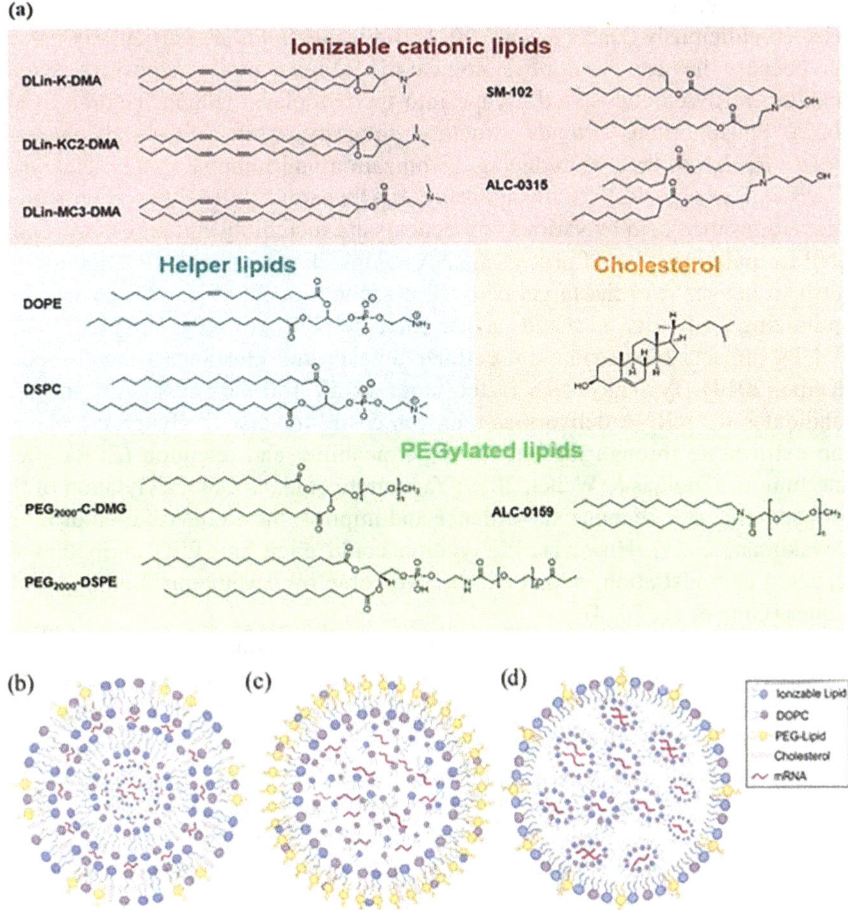

Fig. 1.3 Key design factors of mRNA-LNP formulations. (**a**) LNPs are typically composed of four key lipids: an ionizable lipid, helper phospholipids, cholesterol, and PEG-lipids (**b–d**). Illustrative LNP architectures: (**b**) multilamellar vesicle with concentric lipid bilayers, (**c**) nanostructured core particle, and (**d**) uniform core-shell nanoparticle (Liu et al., 2024)

charge at physiological pH (Adams et al., 2018; Sun & Lu, 2023). The lipids incorporated into LNPs are cholesterol, phospholipids such as Distearoylphosphatidylcholine (DSPC) or Dioleoylphosphatidylethanolamine (DOPE), and polyethylene glycol (PEG) lipids such as Dimyristoylglycerol Polyethylene Glycol 2000 (DMG-PEG2000) (Hao et al., 2023). Moreover, there is evidence that lipid-based carriers exhibit intrinsic immunostimulating activities, and the characteristics of the carrier and lipid composition determine the type and extent of the innate immune response to mRNA vaccines, contributing to vaccine performance and safety (Singha et al., 2018).

In developing LNPs, a delicate balance of multiple parameters is required to ensure efficacy, safety, and stability. Most importantly, the chemical composition of LNPs (e.g., lipids, ionizable lipids, phospholipids, PEG-lipids) should be selected judiciously (Dacoba et al., 2021). Ionizable lipids are particularly important because they can flip at physiological pH from a neutral to a positive charge, enabling mRNA's endosomal escape into the cytoplasm (Ramachandran et al., 2022). Phospholipids provide structural integrity, while PEG-lipids increase blood circulation time by reducing opsonization and immune system clearance (Carissimi et al., 2021; Sathyamoorthy & Dhanaraju, 2016). Since biological interactions mediated by various components are incredibly complex in the body, LNP formulations should protect mRNA against degradation and efficiently target it to deliver it to the target cells. The second aspect of the design involves optimizing the particle size and surface characteristics of LNPs. The particle size of NPs impacts biodistribution, cellular uptake, and clearance from the body (Knipe, 2014). Typically, NPs in the range of 50–150 nanometers are the best candidates for mRNA delivery as they can easily traverse the blood circulation and extravasate through the enhanced permeability and retention (EPR) effect into tumors (Thomas & Weber, 2019; Yamamoto et al., 2022). PEGylation of the surface will evade immune surveillance and improve the biodistribution of LNPs (Jeyakumar, 2021). However, PEGylation could elicit anti-PEG antibodies on repeated administration, which can be overcome by developing shielding techniques (Ginn et al., 2014).

LNP formulations must protect the mRNA from production to administration and delivery to target cells, maintaining consistent encapsulation efficiency and efficient endosomal escape (Hou et al., 2021; Ouranidis et al., 2024). Once the LNPs are taken up by cells via endocytosis, the mRNA must escape from the endosomes into the cytoplasm to be translated into the desired protein, as failure to escape can lead to mRNA degradation within the endosome (Chatterjee et al., 2024). Three fundamental factors influence efficiency: mRNA design, LNP composition, and dose and route of administration. The sequence and structure of the mRNA, including the 5′ and 3′ untranslated regions, coding sequence, and poly(A) tail, affect its stability and translatability (Jia & Qian, 2021). The lipid composition, including helper lipids and PEGylated lipids, affects the stability, release, and targeting of LNPs (Schlich et al., 2021). The amount of mRNA and the method used to deliver it, such as intramuscular or intravenous, influence the distribution and uptake of the vaccine. Perhaps the most crucial element and challenge is that high TE needs to be balanced with minimal, non-cytotoxic side effects. Given the myriad of functional components and molecular interactions required for the effective delivery and dosage of mRNA oncology vaccines, scrupulous design and optimization of these next-generation vaccines are vital.

1.3   mRNA Transfection Efficiency: A Critical Research Problem

One of the most critical bottlenecks in developing mRNA oncology vaccines and nanomedicines in general is the availability of suitable nanocarriers with adequate TE (Huang et al., 2024; Zhou et al., 2024). Although liposomes are currently the most widely used nanocarriers of mRNA vaccines, they can still display cytotoxic properties. It is challenging to search for a safer and more efficient nanocarrier because a suitable nanocarrier must identify materials that effectively combine with drugs, avoid rejection by human cells, possess sufficient delivery capabilities, and be non-cytotoxic (Oladipo et al., 2023). The efficiency of the LNP delivery process depends on several key steps, including delivery to target cells, cellular uptake, endosomal escape, mRNA translation, and immune stimulation (Wang & Yu, 2020). Once at the target cell, LNPs must be efficiently taken up via endocytosis, in which the properties of LNPs, such as size, surface charge, polarization, shape, and lipophilicity, are all influential variables (Seo et al., 2023). LNPs should be readily endocytosed by the target tumor cells to ensure that the mRNA reaches the cell's interior (Fig. 1.4). For the mRNA to function, it must escape from the endosome into the cytoplasm. In this process, ionizable lipids in LNPs can facilitate the escape through the proton sponge effect, which destabilizes the endosomal membrane (Schlich et al., 2021). Once in the cytoplasm, the mRNA must be efficiently translated into protein by the cell's ribosomes, where the stability of the mRNA, its codon optimization, and the effective evasion of immune surveillance mechanisms within the cell are crucial (Chaudhary et al., 2021). In oncology vaccines, the produced protein, i.e., a tumor-specific antigen, must be properly processed and presented on the cell surface to stimulate and effectively elicit an immune response. This involves the antigen-processing pathways of the cell and the activation of T cells.

When it comes to enhancing the delivery mechanisms of these therapeutics, a significant problem remains in predicting and screening the TE of nanocarriers (Huang et al., 2022; Moayedpour et al., 2024; Patel et al., 2023). High TE ensures that antigen-presenting cells can efficiently take up the mRNA-encoded tumor antigens and activate the immune system response against cancerous cells (Gómez-Aguado et al., 2020). This ultimately increases the overall effectiveness of the therapy. However, designing efficient and safe nanocarriers with optimized properties such as adequate transfection efficiency remains a complex challenge. The discovery of optimal nanocarriers and LNP formulations has relied heavily on time-consuming empirical experimentation, which is often fraught with trial and error. Virtual screening, combined with ML, offers a transformative approach to overcome these hurdles. In the context of LNPs for mRNA nanomedicines, virtual screening enables researchers to analyze vast chemical libraries and identify promising candidates based on specific molecular descriptors, i.e., molecular fingerprints. This in silico approach can significantly reduce the time and costs associated with experimental trials, allowing focus to be placed on the most promising candidates for development. Virtual screening can rapidly evaluate how different lipid compositions and formulations impact biological functionality, providing a first-pass filter that narrows down the options for subsequent experimental validation (Oršolić, 2023). While traditional virtual screening methods rely on molecular dynamics simulations or rule-based models, ML can dramatically enhance the prediction of LNP performance with greater speed and accuracy (Sakiyama, 2009). By training on datasets containing known LNP structures and their corresponding biological outcomes, ML algorithms can learn complex patterns and relationships that may be difficult to capture with traditional computational approaches. ML handles

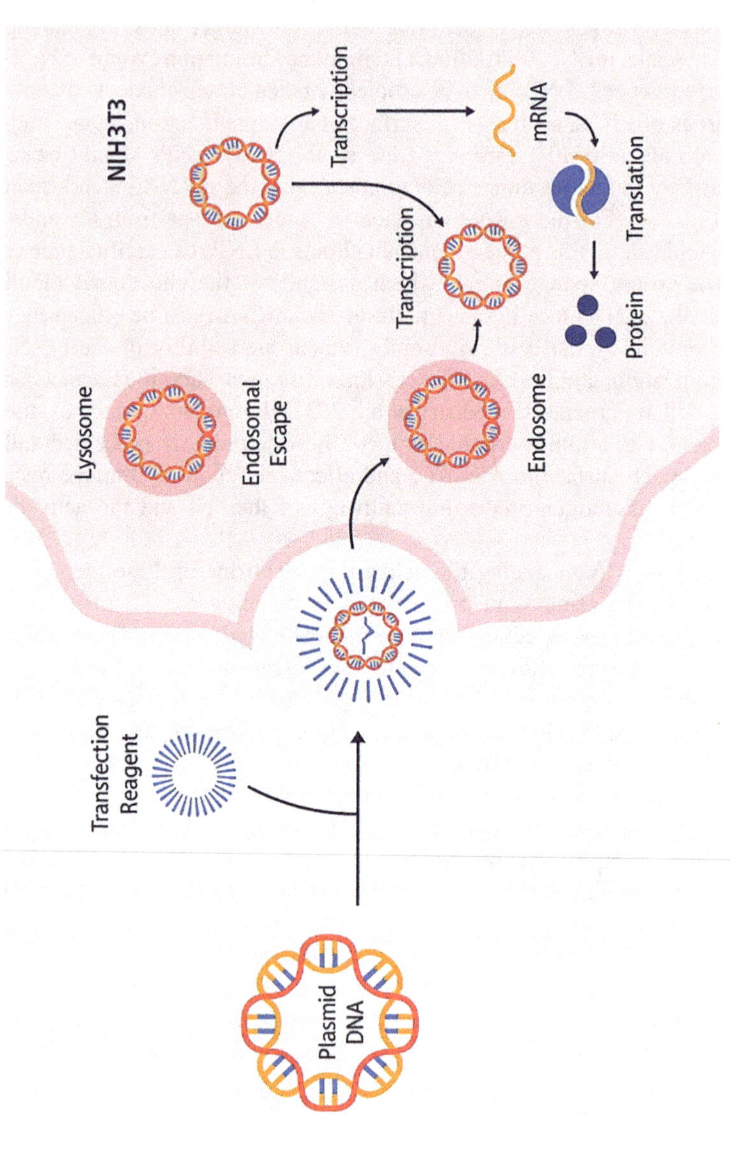

Fig. 1.4 mRNA transfection mechanistic diagram (Verma, 2023)

high-dimensional data, making it ideal for LNP design, where multiple molecular features are considered simultaneously (Carter, 2023). Once trained, ML models can rapidly screen new LNP candidates by predicting critical outcomes such as transfection efficiency, stability, and biocompatibility. This accelerates the in silico screening process, allowing for more precise optimization and identifying formulations that balance efficacy and safety.

Given the critical role of TE in mRNA oncology vaccines and nanomedicines, this work focuses on optimizing the in silico prediction of LNP TE using ML combined with advanced chemoinformatic techniques, including molecular fingerprinting, to accelerate the design and development of next-generation nanomedicines. ML has opened the door to the rapid design and identification of safer, more efficient nanocarriers, thereby improving the efficacy and translatability of next-generation mRNA cancer vaccines (Ding et al., 2023; Öztürk et al., 2018). LNP design is critical for effective mRNA vaccine delivery, as LNPs must be stable and biocompatible with the proper mRNA encapsulation and TE profile. Previous studies have used ML for virtual screening. Still, these models are often hindered by poor/limited data quality, data imbalances, instability, scalability issues, and a poor overall fit, given the complexity of biological data (Mey et al., 2021). Ding et al. (2023), Obaido et al. (2024), Mey et al. (2021), and Öztürk et al. (2018) have highlighted the need for more sophisticated models that can effectively handle the complexity of biological systems, chemical structures, and the multifaceted nature of nanomedicine. Integrating advanced ML techniques, such as semi-supervised learning and pseudo-labeling, as well as synthetic data augmentation, can significantly enhance the predictive modeling and in silico screening of LNPs for nanomedicine design and development. Furthermore, incorporating molecular fingerprinting provides a detailed understanding of the molecular characteristics that influence TE, which can be pivotal in designing more effective nanocarriers. This research hypothesizes that integrating advanced ML techniques with chemoinformatics, including molecular fingerprinting, pseudo-labeling, and synthetic data augmentation, can enhance the prediction accuracy and in silico screening of LNP transfection efficiency for mRNA oncology vaccines, thereby accelerating the design and development of next-generation nanomedicines. Building on the insights gained from the existing literature, this research integrates a methodological approach that leverages ML to address the identified gaps. The following methodology section outlines the experimental setup and the data-driven techniques employed to address this research problem.

Chapter 2
Machine-Learning Enhanced In Silico Screening: A Methodological Approach

Designing LNPs that can actively function within an in vivo biological system is a cornerstone of nanomedicine development. LNP nanocarriers should be stable, biocompatible, have a specific size, and effectively encapsulate, transfect, and release the mRNA (Hou et al., 2021). ML models can predict the properties of LNPs based on various parameters, aiding in the optimization of their design and enhancing the process of designing LNPs for delivering mRNA cancer vaccine nanomedicines that encode TAAs or neoantigens. Optimizing the design of LNPs to maximize their transfection efficiency and, subsequently, the efficacy of mRNA vaccines remains a complex and challenging task (Weng et al., 2020). Traditional laboratory-based approaches to optimizing LNP design often involve labor-intensive and time-consuming experimental procedures. Laboratory researchers and bench scientists must often test many formulations, varying the types and ratios of lipids, to identify effective combinations (Evers et al., 2018). This trial-and-error method is inefficient and limited in exploring the massive chemoinformatic catalogs of potential LNP formulations. ML is a subfield of artificial intelligence (AI) that employs algorithms for the analysis of complex input data to make predictions or decisions without being explicitly programmed for every input-output pair (Mey et al., 2021). ML determines patterns in data by identifying hidden feature patterns and applying them to use contexts without explicitly programmed instructions, and is a fundamental process in data mining, pattern recognition, and predictive analysis. The application of ML in biotechnology is vast, from finding drug targets in genomic data to designing de novo proteins, and it can accelerate development by reducing costs and increasing precision in the handling of biotechnological data. ML enables the high-throughput screening of molecular compounds and formulations for relevant molecular feature interactions, uncovering unique patterns and relationships that might not be apparent through traditional analysis methods (Bannigan et al., 2021). These insights can then be used to predict the performance of new formulations, guide the design process, and accelerate the development of potential therapeutic

K. W. Ramadurai, A. Banerjee, *Machine Learning-Driven Rational Design in Nanomedicine*, SpringerBriefs in Bioengineering, https://doi.org/10.1007/978-3-032-04012-1_2

candidates. Virtual screening, combined with ML, offers a practical approach to overcome these hurdles by leveraging computational models to predict LNPs' biological activity and properties before being synthesized and tested in the laboratory (Fig. 2.1) (Santana et al., 2021).

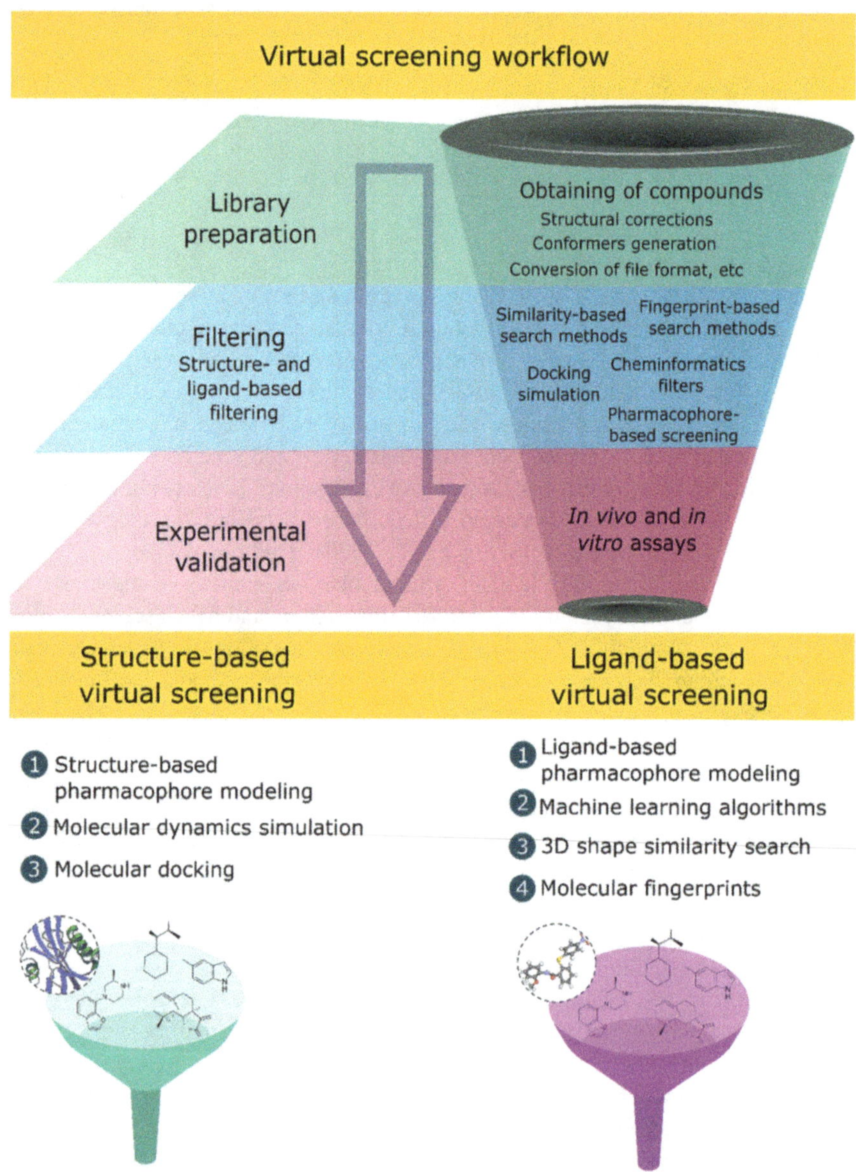

Fig. 2.1 Virtual screening computational methods, including ligand-based virtual screening and structure-based virtual screening approaches, are used to identify novel compounds (Santana et al., 2021)

As will be thoroughly discussed later in this work, ML models can predict which combinations of lipids are likely to result in high transfection efficiency based on their molecular fingerprints. These predictions can then inform the virtual screening process, allowing researchers to prioritize the most promising candidates for experimental testing. Moreover, the iterative nature of machine learning enables the continuous refinement of models as new data becomes available, thereby improving their predictive accuracy over time. In silico models can simulate numerous variations of LNP formulations, evaluate their potential efficacy, and predict their behavior in biological systems, all before any wet-lab experimentation (Fig. 2.2). ML-accelerated virtual screening can also enable the discovery of novel LNP formulations that traditional experimental protocols may have overlooked by uncovering unique and subtle molecular interactions.

Traditional learning approaches employed in previous literature require extensive labeled datasets, which are scarce and often encounter difficulties when modeling biological data, including model instability and data imbalances, leading to poor model performance (Mey et al., 2021; Vercio et al., 2020). These datasets often need better molecular characterization and labeling domains, thus necessitating advanced chemoinformatic integration for effective, high-throughput virtual screening. Further integration of approaches such as semi-supervised machine learning (SSL) can effectively leverage a small amount of labeled data with a larger volume of

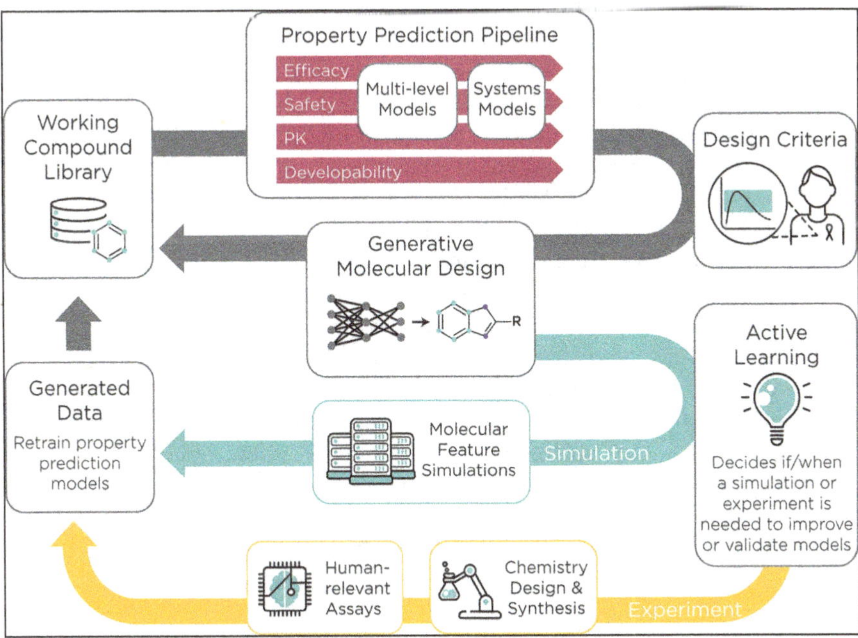

Fig. 2.2 Biological data processing and an integrative workflow combining ML and biological data processing for in silico drug screening. The process initiates with a compound library undergoing ML-driven property prediction via multi-level and systems models, informed by defined biological design criteria (Hinkson et al., 2020)

unlabeled data to uncover and extrapolate novel feature correlations that can be applied to massive volumes of biological data. In predicting LNP TE, SSL can harness the vast amounts of unlabeled biological data, integrating it with sparse labeled datasets to train more accurate predictive models. In dealing with the complex nature of biological interactions and delivery mechanics behind mRNA vaccine delivery, a refined approach is required beyond basic learning approaches; thus, this chapter explores an integrated approach. This chapter examines the conceptual application of supervised and semi-supervised learning approaches with advanced techniques, including iterative pseudo-labeling, combined with molecular fingerprinting, to enhance the prediction of LNP TE. This will enable the design of better and high-throughput in silico screening of viable LNPs for mRNA vaccines, thereby bridging the gap in existing methodologies.

2.1   Machine Learning & Molecular Fingerprinting

Despite the advancements in LNP design technology and ML applications in biotechnology research, a significant gap remains in predicting LNP TE (Yang et al., 2022). Previous studies and literature have explored the use of basic supervised learning, whereby models are trained on limited labeled datasets (Cheng et al., 2023; Ding et al., 2023; Li et al., 2024; Maharjan et al., 2024; Wang et al., 2022). However, the effectiveness of supervised learning is contingent upon the availability of large, high-quality labeled datasets that often lack sufficient structure-activity modeling and are scarce due to the high costs and logistical challenges associated with experimental data collection (Obaido et al., 2024). This challenge highlights the importance of exploring additional alternative ML approaches that can leverage unlabeled biological data. Because it simultaneously utilizes labeled and unlabeled data, SSL lies between supervised and unsupervised learning and is highly applicable in biomedical research, where labeled data are often scarce but large amounts of unlabeled data are routinely generated. One promising SSL technique is pseudo-labeling, where a model trained on labeled data predicts labels for the unlabeled data (Fig. 2.3). These predicted labels, or pseudo-labels, are then combined with the labeled data to retrain the model, effectively increasing the amount of labeled data (Majchrowska et al., 2021). By incorporating pseudo-labeling, one can effectively leverage a larger pool of unlabeled data, which can be utilized to retrain the initial supervised model and further enhance predictive accuracy. Ideally, these pseudo-label-enhanced models can identify promising new formulations more quickly and provide deeper functional insights into the molecular features that drive high TE, thus reducing the need for lengthy experimental bench testing and accelerating preclinical therapy development.

While the functional premise of these ML algorithms holds promise in predicting and interpreting complex bioactivity data for developing mRNA vaccines, it is crucial that structure-activity modeling is conducted (Kwon et al., 2022). This is where the introduction of molecular fingerprinting comes into play. The molecular

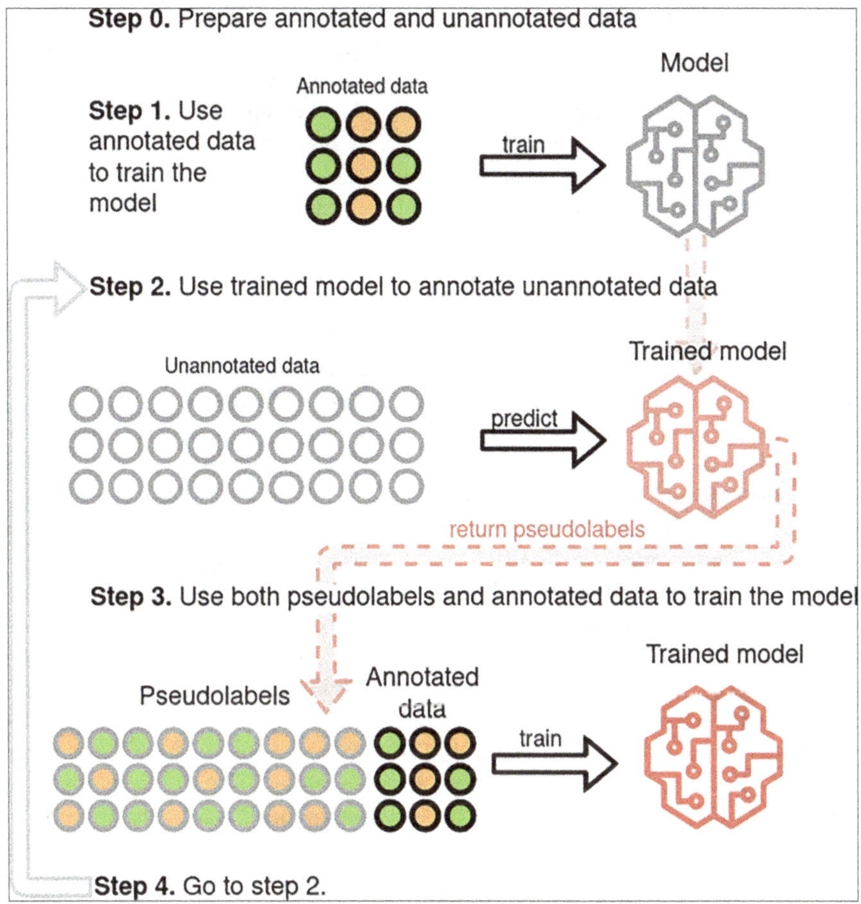

Fig. 2.3 SSL with Pseudo-Labeling (Majchrowska et al., 2021)

fingerprinting process can leverage the molecular structure information encoded in the simplified molecular-input line-entry system (SMILES) string that describes the chemical structure of a molecule (Yang et al., 2022). Molecular fingerprinting involves manipulating molecules based on their physicochemical properties, structural features, and biological activities (Yang et al., 2022). In the context of LNP-mRNA vaccines, molecular fingerprinting can generate detailed profiles of various lipid components and their interactions with mRNA. These fingerprints include data on lipid composition, particle size, surface charge, and other critical molecular parameters that influence the performance of LNPs. By creating a comprehensive database of molecular fingerprints, researchers can systematically study how different formulations affect mRNA vaccine stability, delivery efficiency, and immunogenicity (Vora et al., 2023). SSL enables the use of a smaller labeled dataset in conjunction with a larger pool of unlabeled data, thereby amplifying the learning process and allowing the model to generalize more effectively and make more accurate predictions. Bench scientists and researchers can begin with a limited set of experimental results, i.e., labeled data, and expand their model's training dataset by predicting labels for the expansive unlabeled data generated through molecular fingerprinting. The integration of SSL with molecular fingerprinting can improve the design of LNP-mRNA oncology vaccines by predicting which lipid combinations will most effectively encapsulate mRNA and ensure its stability based on the molecular fingerprints of known successful formulations. Additionally, SSL can help uncover subtle relationships between molecular characteristics and biological outcomes, enabling the creation of LNPs tailored for specific oncological target domains. By analyzing the molecular fingerprints of lipids that demonstrate high efficacy in delivering mRNA to specific cancer cell types, SSL can aid in designing LNPs optimized for different tumor microenvironments, which are often heterogeneous across patients and challenging to tailor effective therapies for.

To effectively screen and design better LNPs using ML approaches, the data of these molecules must be extrapolated, translated, and analyzed through a process known as molecular fingerprinting. This process can translate LNP chemoinformatic data into a format that can be analyzed by ML algorithms by capturing each molecule's unique structural and chemical characteristics systematically (Wang et al., 2022). A specific type of fingerprinting, known as Morgan molecular fingerprinting or circular fingerprinting, is utilized to enhance the LNP predictive modeling capabilities for ML applications (Capecchi et al., 2020). This technique generates a compact fingerprint of a molecule by encoding its molecular structure into a fixed-length binary vector, where each bit represents the presence or absence of a particular substructure within the molecule (Fig. 2.4) (Rogers & Hahn, 2010). The process begins by considering each atom in the molecule and its immediate environment, forming an initial set of substructures. The algorithm then expands these substructures by including atoms in successive connectivity layers until a predefined radius is reached (Rogers & Hahn, 2010). Each unique substructure is hashed to a specific

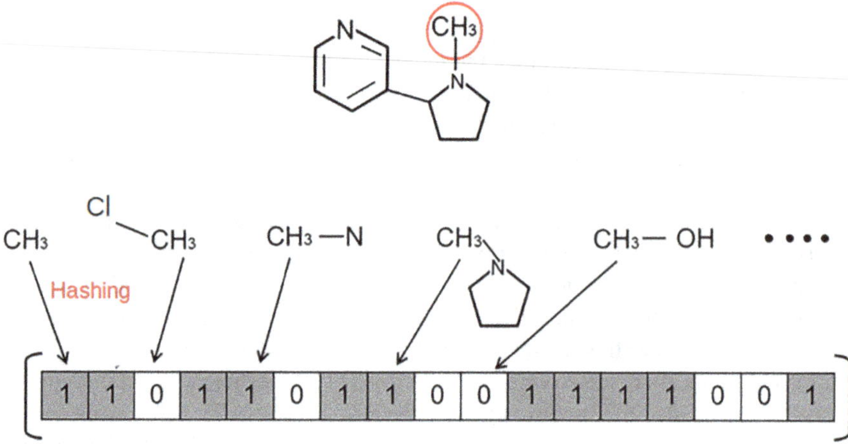

Fig. 2.4 Molecular fingerprint schematic (Mendolia et al., 2020)

position in the fingerprint vector, which is then set to 1 if that substructure is present in the molecule (Capecchi et al., 2020; Rogers & Hahn, 2010). Unlike other simple structural descriptors, Morgan fingerprints capture the presence of substructures, functional groups, and specific atom environments within a molecule (Yang et al., 2022). Given that Morgan molecular fingerprinting can capture and analyze complex chemical properties of lipids and convert these molecular structures into fingerprints, this can facilitate the identification of structural motifs that correlate with desirable properties, such as high TE (Yang et al., 2022). This ability to capture both the local chemical environments and the overall molecular topology of LNP molecules enables the systematic exploration and optimization of formulations and structural conformations, making it ideal for therapeutic design and molecular modeling applications within complex biological systems (Cheung et al., 2022; Ramezanpour, 2019).

2.2   Chemoinformatic-Based Molecular Feature Extraction & Processing

The ability to extract molecular features and create molecular fingerprints provides the necessary foundation and data for developing ML models to predict LNP TE effectively. Two essential tools in this feature extraction process are the RDKit and pandas, which provide the capability to generate, organize, and preprocess molecular data. RDKit is a toolkit for working with chemical informatics data that can facilitate the transcription of SMILES strings into molecular features (Badie-Modiri & Kivelä, 2023). It is widely used for in silico drug discovery and development, and it plays a pivotal role in molecular fingerprinting for LNP-based mRNA vaccines. RDKit can calculate a wide array of molecular descriptors and numerical values representing various chemical properties of molecules. These descriptors are essential for characterizing the features of LNPs, allowing RDKit to generate molecular fingerprints — bit strings representing the presence or absence of specific substructures within a molecule. These fingerprints serve as input features for ML models (Fig. 2.5) (Shin, 2023).

The specific types of molecular descriptors that RDKit can generate include molecular weight, the number of rings, logP (partition coefficient), polar surface area, and the number of hydrogen bond donors and acceptors, which are critical for understanding the behavior of LNPs in biological systems. RDKit enables the manipulation and visualization of chemical structures, which is useful for modifying LNP formulations and examining how changes in structure affect TE. RDKit integrates with ML libraries, enabling the use of extracted molecular features as input for predictive models (Wang et al., 2022). The detailed code outline of these molecular feature extraction techniques is described in Appendix A.

Another important tool is Pandas, a Python library critical in managing the diverse data generated during in silico screening, which can efficiently organize, clean, and preprocess the molecular data generated by RDKit and create databases

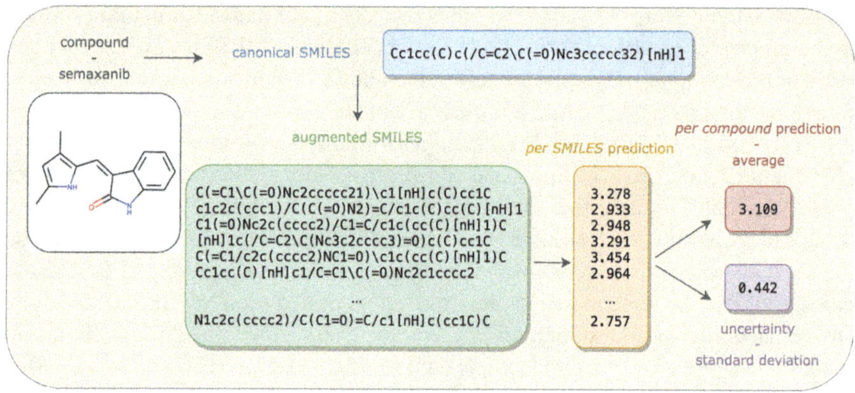

Fig. 2.5 Molecular fingerprinting within RDKit in Python (Kimber et al., 2021)

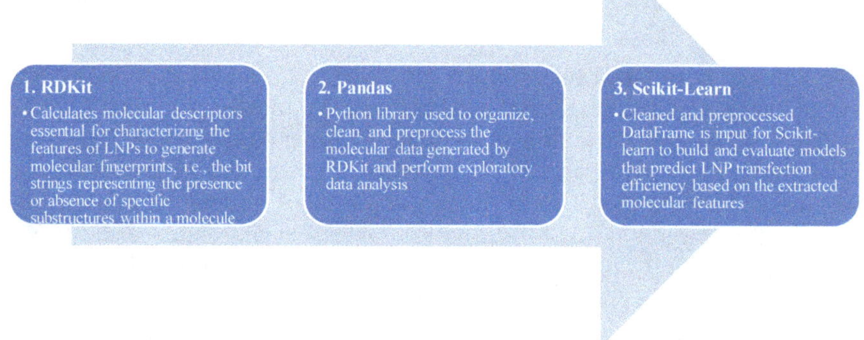

Fig. 2.6 RDKit, Pandas, and Scikit-Learn integrated workflow

that include molecular fingerprints, experimental outcomes, and predicted properties (Lindley et al., 2023). Pandas DataFrames store and manage tabular data, with each row in a DataFrame representing a different LNP formulation, while each column can represent a different molecular descriptor or fingerprint (Lindley et al., 2023). Pandas can also facilitate exploratory data analysis (EDA), providing insight into the dataset's underlying descriptive statistics, patterns, and relationships (Sahoo et al., 2019). Pandas also integrates with scientific libraries such as NumPy, Matplotlib, and Scikit-learn, providing a streamlined workflow from data extraction to model development and evaluation (Fig. 2.6). The cleaned and preprocessed DataFrame created in Pandas is used as input for ML models with the Python library tool Scikit-learn. Scikit-learn enables ML in Python to build and evaluate models that predict LNP transfection efficiency based on the extracted molecular features and ensure that the data is organized and ready to be codified to enable adequate predictive modeling.

2.3 Advanced ML Techniques and Molecular Fingerprinting: Optimizing In Silico Modeling

With the foundation set with data cleaning, processing, and formatting, advanced ML techniques can be developed and applied. In biotechnology research, pseudo-labeling is a unique approach that can further enhance the in silico screening of de novo nanomedicines. Pseudo-labeling involves using a trained ML model to generate labels for unlabeled data, creating a richer dataset for further training. This approach can accelerate in silico screening and modeling by leveraging unlabeled experimental data that would otherwise remain underutilized. This iterative process can enhance the efficiency of identifying optimal LNP formulations by predicting which lipid combinations will yield stable and efficient mRNA encapsulation, reduce toxicity, and enhance targeted delivery (Tilstra et al., 2023). Pseudo-labeling can allow researchers to explore a broader range of formulations and conditions than would be feasible through experimental testing alone and identify critical parameters that influence the performance of LNP-mRNA vaccines. Specifically, as the model iteratively learns from labeled and pseudo-labeled data, it can uncover complex relationships between molecular structures and their biological effects. This deeper understanding can lead to more precise engineering of LNPs for effective in vivo mRNA delivery.

Drug development workflows can finally transition from an experimental approach to a data-informed predictive methodology using pseudo-labeling. This approach can optimize several critical aspects of the design of the LNPs, which involves selecting and tuning various lipid components for optimal delivery and immune response. This includes predicting and validating the efficacy of different lipid formulations on a small scale, which can then be applied to a broader dataset with iterative refinement. The pseudo-labeling process comprises several critical steps: initial model training, label prediction, data augmentation, and model retraining (Fig. 2.7). Initial model training begins with training a supervised learning model on the available labeled data. This model serves as the foundation for

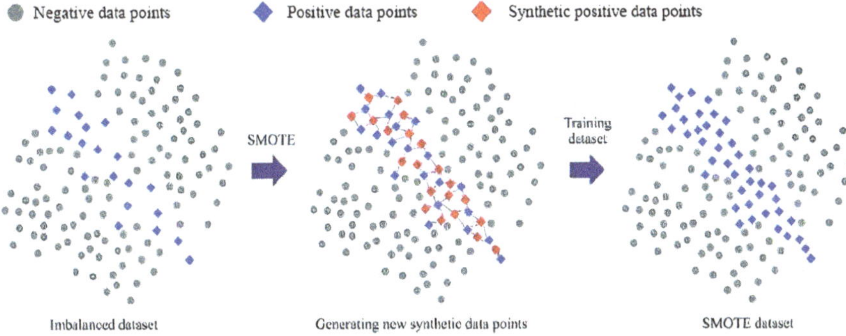

Fig. 2.7 SMOTE technique process (Hairani et al., 2023)

generating pseudo-labels. Next comes label prediction, in which the trained model predicts labels for the unlabeled data. This is followed by data augmentation, in which the pseudo-labeled data is combined with the original labeled data, creating an enriched dataset that includes actual and predicted labels (Yu et al., 2022). The model is then retrained on this augmented dataset, enabling it to learn from a more comprehensive set of examples and thereby improve its generalization capability (DeVries & Taylor, 2017).

The predicted labels, or pseudo-labels, are then treated as if they were true labels, and the model is retrained using both the original labeled data and the newly pseudo-labeled data (Yu et al., 2022). This iterative process amplifies the training data, allowing the model to learn more nuanced patterns and make more accurate predictions. This approach can predict the properties and efficacy of various lipid formulations without requiring extensive experimental validation for each proposed candidate. The optimization process begins with a small dataset of known LNP formulations, where each formulation's stability, encapsulation efficiency, transfection efficiency, cellular uptake, and immunogenicity performance are well-characterized and in vitro/in vivo validated. For the sake of this research, the focus is placed on TE, given the availability of data, on which the model can be initially trained to understand the relationships between lipid composition, molecular features, and efficacy. Once trained, the model generates pseudo-labels for a much larger set of unlabeled LNP formulations. These pseudo-labels provide a preliminary assessment of how each unlabeled formulation might perform, effectively expanding the dataset and enabling better model training. The model, enhanced by the pseudo-labels, can identify promising candidates that exhibit high stability and delivery and uncover complex interactions between different lipid components and their collective impact on the LNP's performance, which might be challenging to discern through traditional experimental approaches (Mitra, 2011).

One of the most significant issues when developing advanced ML models is imbalanced datasets, whereby the number of samples in one class significantly outweighs the other and causes biases that lead to the model performing poorly on the minority sample class (Hairani et al., 2023). This issue was present in the research, as the number of samples in the positive TE class exceeded the number of samples in the negative TE class, which is addressed in the next chapter. The Synthetic Minority Oversampling Technique (SMOTE) addresses this data imbalance issue and enhances prediction accuracy and outcomes (Fig. 2.7).

SMOTE generates synthetic samples for the minority sample class and balances the dataset. These synthetic samples are created by interpolating between existing minority class instances, where a sample from the minority class is randomly selected to effectively find its k-nearest neighbors (Tarawneh et al., 2020). Then, it generates new samples along the line segments, joining the original sample and its neighbors, creating a more balanced dataset without duplicating existing minority samples, which can lead to overfitting (Tarawneh et al., 2020). In this research, SMOTE was applied to balance the dataset of LNP TE. Initially, the dataset had an imbalance between effective and ineffective transfection cases. By applying SMOTE, synthetic instances were generated of the minority class, i.e., ineffective transfection cases, to match the number of instances in the majority class, i.e., effective transfection cases. This optimized process accelerates the iterative design, testing, and refinement cycle in the preclinical development of nanomedicine. While pseudo-labeling and SMOTE offer substantial benefits, they are not without challenges. The quality of the pseudo-labels depends on the accuracy of the initial model. If the initial model's predictions are significantly off, the pseudo-labels can introduce noise into the dataset and negatively impact performance (Lokhande et al., 2020). Confidence thresholds were used to counter this by using only the most confidently predicted pseudo-labels for retraining.

Furthermore, data sparsity and ensuring that the model generalizes well to unseen data are critical, as the model should not be overly reliant on the synthetic samples generated, which are unlikely to emulate real-world data perfectly (Butcher et al., 2021). Thus, additional techniques, such as gradient boosting via XGBoost and hyperparameter tuning via GridSearchCV, can be utilized and explored later in this work. These techniques can refine and enhance model prediction by sequentially building a series of decision trees, where each tree corrects errors made by the previous ones (Mitchell & Frank, 2017). Furthermore, XGBoost can prevent overfitting by integrating regularization and handling missing data efficiently. Thus, for this research, XGBoost's ability to handle complex data patterns and improve classification precision and recall makes it an ideal choice for integrating into the advanced SSL techniques developed for predicting LNP TE. The proper integration of molecular fingerprinting to define molecular features and advanced machine learning models to analyze chemoinformatic data presents an innovative strategy for further optimizing the rational design of next-generation nanomedicines. Given the development of a methodological framework using advanced ML, the next step involves evaluating how these methods perform in practice. The following two chapters explore the models' results and outcomes, highlighting key performance metrics and their implications for screening optimization for TE.

Chapter 3
Supervised Machine Learning Implementation & Results

Developing an effective ML model for predicting LNP TE began with implementing a Random Forest Classifier (RFC). This chapter outlines the step-by-step process from data preparation to model evaluation, highlighting the unique insights and challenges encountered along the way. The first and foundational step was preparing the dataset. The starting point involved the procurement and curation of a dataset of over 600 LNPs along with their level of transfection efficiency (TE), which was conducted after extensive literature analysis, as there are minimal public data resources that can be utilized to develop ML models for LNP design (Ding et al., 2023; Zhang et al., 2022). Specifically, the initial dataset was procured via Ding et al. (2023), which provided publicly available open-source data for research purposes. The chemical structures of the LNPs are represented as SMILES string codes followed by their TE, which is the target variable of interest, and coded into a binary label: those with satisfying TE are represented as a 1, and those with unsatisfying TE are defined as a 0. This categorization was done to facilitate model development on an integrated dataset. Each LNP record included TE outcomes and 11 chemoinformatic and molecular features describing the properties of the LNPs. These features spanned chemical composition, physical properties, and experimental conditions, providing a comprehensive dataset for training ML models to predict TE accurately. The SMILES strings were concatenated to form a combined representation of the LNPs, treating the entire LNP as a single molecule for molecular descriptor calculations. The SMILES strings contain the chemical structures of the four components of the LNPs, which include an ionizable or cationic lipid for complexing the polyanionic RNA, a helper phospholipid to stabilize the LNP, sterols for facilitating endosomal escape, and lipid-anchored

K. W. Ramadurai, A. Banerjee, *Machine Learning-Driven Rational Design in Nanomedicine*, SpringerBriefs in Bioengineering, https://doi.org/10.1007/978-3-032-04012-1_3

poly(ethylene glycol) (PEG) lipids for increasing circulation time (Puccetti et al., 2023; Shi et al., 2024; Swetha et al., 2023). The concatenated LNP SMILE strings were then converted into Extended-Connectivity Fingerprints (ECFPs), which serve as comprehensive, binary representations of molecular structures for input into ML models utilizing RDKit (Wigh et al., 2022).

The Pandas library in Python was used to load the data into a DataFrame, which provided a structured way to manage and manipulate large volumes of information. This step was crucial for organizing the data and analysis. Exploratory data analysis (EDA) was conducted to examine the data's structure and identify any anomalies or patterns. Functions such as data.info() and data.describe() were used to assess the data types and discover missing values. Handling missing values is a crucial aspect of data preparation, as these gaps can significantly impact the model's performance. Columns with missing values were identified, and missing entries were filled out with mean or median values. When entire columns contained substantial amounts of missing data, they were entirely removed to avoid introducing bias. Relevant molecular feature selection was conducted to refine the dataset, with redundant or irrelevant data removed for model integrity (Cánovas-Segura et al., 2019). This included encoding categorical variables through one-hot encoding, which allowed these variables to be effectively utilized in the ML algorithms (Seger, 2018). The features were standardized using StandardScaler from scikit-learn to transform them to a standard scale, ensuring that each feature contributed equally to the model's performance and preventing any single feature from dominating due to differences in scale.

Once the data was prepared, it was split into training and testing sets to evaluate the model's performance. This split was performed using the train_test_split function, where 80% of the data was allocated for training and 20% for testing. This partitioning was necessary for assessing the model's ability to generalize to unseen data, providing a fair evaluation of its capabilities (Niarman & Amuharnis, 2023). With the data prepared, the initial Random Forest Classifier was trained. The RFC was selected for its effectiveness in handling complex datasets, where the model was trained on labeled training data using the RandomForestClassifier from scikit-learn. This step involved setting up the classifier with 100 trees or estimators, which allowed the model to aggregate predictions from multiple decision trees, thus enhancing accuracy and resilience against overfitting (Mosavi et al., 2020). These steps, including molecular feature extraction, RFC development, and the SSL model, will all be broken down into a chronological sequence throughout this chapter, as shown in Fig. 3.1.

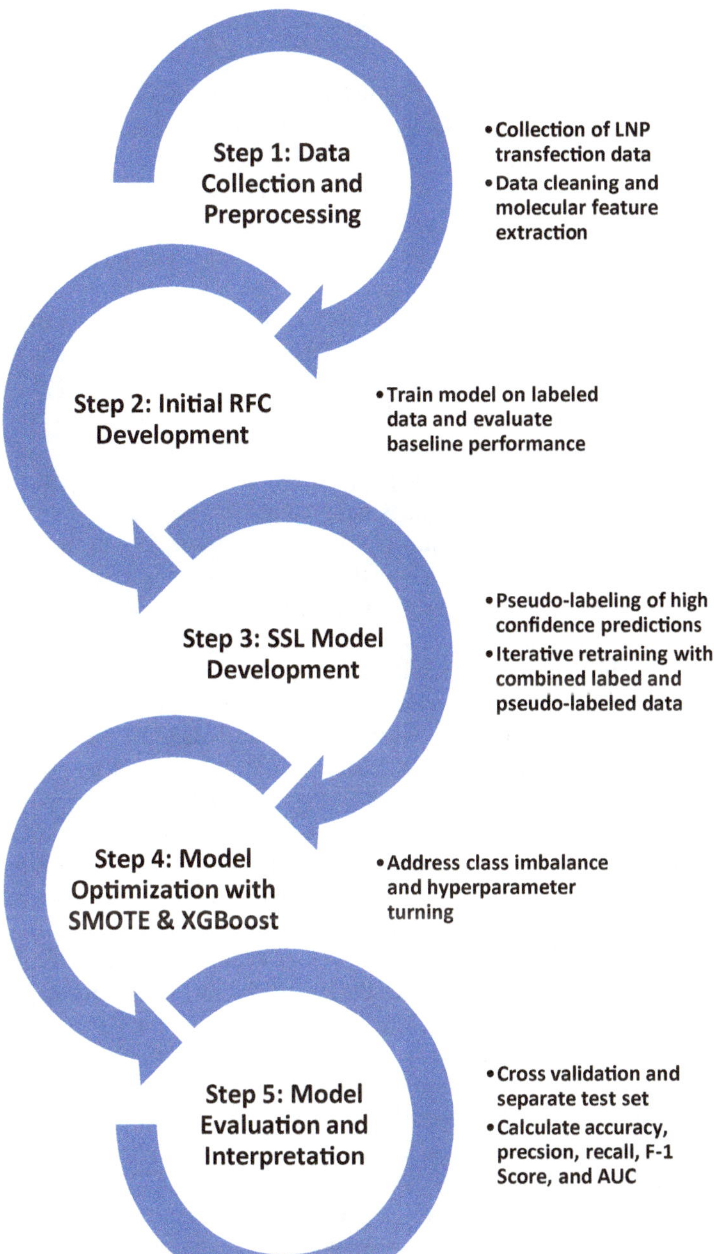

Fig. 3.1 Research process methodology and workflow

3.1　Molecular Feature Processing & Analysis Framework

Once the dataset was concatenated, utilizing RDKit in Python, 11 molecular features were calculated, including molecular weight, number of rings, logP, polar surface area (PSA), unsaturation (hydrogen binding), ionizability, presence of a helper lipid, pegylation, the number of hydrogen bond donors and acceptors, and the number of rotatable bonds present (S'ari et al., 2018). These molecular features ensure efficiency and effectiveness when designing LNPs for mRNA nanomedicines. Molecular weight is essential as it influences the LNP's size, stability, and ability to encapsulate mRNA (Schoenmaker et al., 2021). The number of rings can affect the rigidity and packing density of the lipid bilayer, impacting the LNP's structural integrity and release profile (Alipour et al., 2017). LogP, a measure of a lipid's hydrophobicity, determines its cellular uptake and endosomal escape abilities (Peetla et al., 2014).

Polar surface area (PSA) is crucial for solubility and bioavailability, as it directly influences the interaction of liposomes with aqueous biological environments and cellular membranes (Alfagih et al., 2020). Unsaturation refers to the number of double bonds that the lipid chains contain, imparting fluidity to the lipid bilayer, which is vital for the fusion of the LNP with cellular membranes and the release of mRNA (Tenchov et al., 2021). Ionizable lipids are designed to be neutral at physiological pH but become positively charged in acidic environments, thus aiding the escape of mRNA from endosomes into the cytoplasm (Ramachandran et al., 2022). Helper lipids, such as cholesterol, are incorporated to stabilize the LNP structure and enhance cellular uptake (Cheng & Lee, 2016). Pegylation, the addition of polyethylene glycol (PEG) chains, provides a hydrophilic coating that extends the circulation time in the bloodstream by reducing opsonization and immune system clearance (Hoang Thi et al., 2020). The number of hydrogen bond donors and acceptors in lipid molecules influences their interaction with water and other biological molecules, thereby affecting the stability and delivery efficiency of LNP (Cornebise et al., 2022). Finally, the number of rotatable bonds affects the flexibility of lipid molecules, which can influence the overall fluidity and fusion capability of the LNP with cellular membranes, thereby ensuring the effective delivery of the mRNA payload (Gareev et al., 2023). These unique features must be carefully optimized to create effective LNPs to drive high TE, balancing stability, delivery efficiency, and biocompatibility. Once the features were created, several descriptive statistics analysis frameworks were conducted, including a correlation matrix, histograms, and a Mann-Whitney Test, described in Fig. 3.2 and Table 3.1.

The correlation matrix provided insight into the relationships between TE and various molecular features, as well as the interrelationships among these molecular features. logP showed the strongest positive correlation with transfection efficiency ($r = 0.51$). This suggests that higher logP values, which indicate higher lipid solubility, are associated with higher TE. Molecular Weight correlated moderately ($r = 0.32$) with TE, the Number of Rotatable Bonds correlated moderately positively ($r = 0.31$), the Number of Hydrogen Acceptors correlated weakly ($r = 0.15$), the

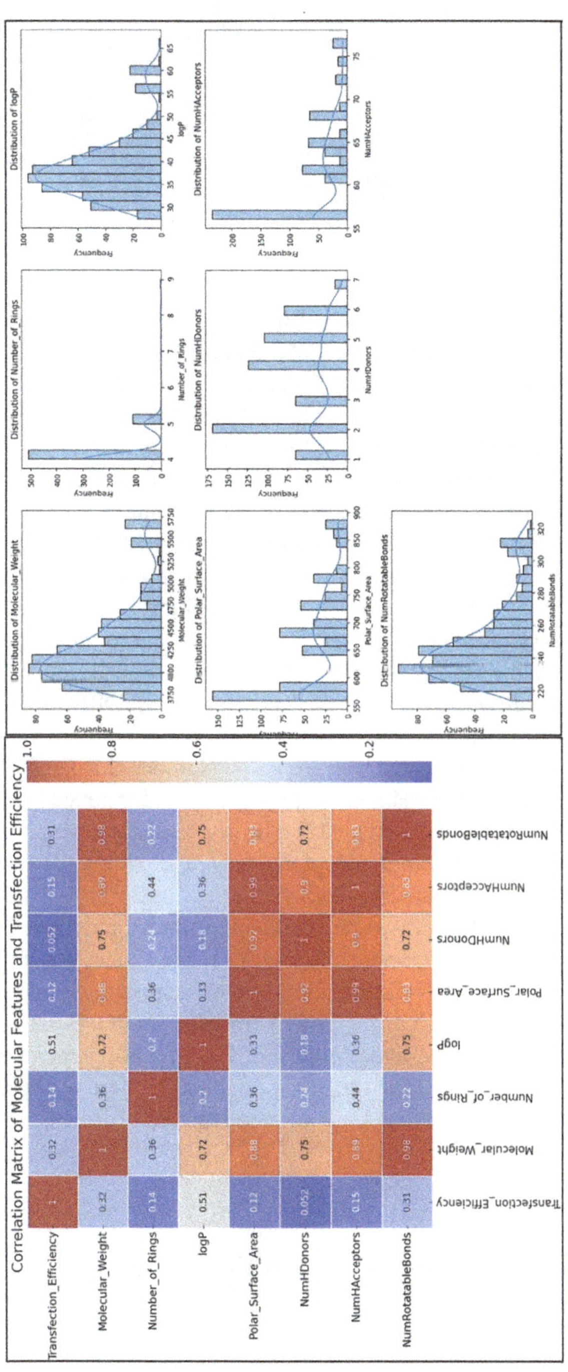

Fig. 3.2 Correlation matrix and histogram of calculated molecular features

Table 3.1 Mann-Whitney test results

Molecular Feature	U-Statistic	p-Value
Molecular_Weight	26136.5	2.66E-04
Number_of_Rings	25816.0	1.4E-08
logP	13996.0	1.66E-24
Polar_Surface_Area	32799.0	9.68E-01
NumHDonors	31689.5	5.14E-01
NumHAcceptors	30999.5	2.96E-01
NumRotatableBonds	24151.5	2.35E-06

Number of Rings correlated weakly (r = 0.14) and Polar Surface Area (r = 0.12), and Number of Hydrogen Donors correlated very weakly (r = 0.05). These high inter-correlations between MW, PSA, and hydrogen acceptor/donor counts indicate that these features primarily reflect overall molecular size or complexity, and collectively, they can influence TE. There were several other sets of intercorrelations, including the high correlation between Molecular Weight and the number of rotatable bonds (r = 0.98), the number of hydrogen acceptors (r = 0.89), and the polar surface area (r = 0.88). Polar Surface Area correlated highly with number of hydrogen acceptors (r = 0.99) and number of hydrogen donors (r = 0.92), logP correlated moderately with molecular weight (r = 0.72), number of rotatable bonds (r = 0.75), and number of hydrogen acceptors (r = 0.36), and Number of Rings correlated moderately with polar surface area (r = 0.36) and number of hydrogen acceptors (r = 0.44). Understanding the positive correlations between features such as molecular weight, log P, and polar Surface area can help design LNPs that balance these properties for optimal performance.

Molecular Weight, Number of Rings, logP, and Number of Rotatable Bonds have statistically significant p-values, indicating a significant difference in their distributions between the two groups based on TE. The molecular features, including PSA, number of hydrogen donors, and acceptors, do not show significant differences between the groups, as indicated by their higher p-values. The Mann-Whitney U test was used to assess whether two independent samples came from the same distribution, without assuming normality (Table 3.1). Molecular Weight, Number of Rings, logP, and Number of Rotatable Bonds show statistically significant differences between the groups with different TEs (p-value <0.05). Polar Surface Area, Number of Hydrogen Donors, and Number of Hydrogen Acceptors do not show statistically significant differences, i.e., a p-value >0.05. These results indicate that certain molecular features, such as logP and the Number of Rings, are likely more influential in determining TE.

3.2 Random Forest Classifier with Five-Fold Cross Validation

A Random Forest classifier (RFC) was utilized for the TE analysis as it was ideal for handling numerical and categorical features, does not require the assumption of linear relationships, and provides a straightforward way to assess feature importance (Ouma et al., 2024). In deploying this framework, the dataset was first processed and normalized, with the SMILE feature variables represented as numerical values and the categorical variables represented as binary values. Normalization and balancing of features were performed using the StandardScaler from scikit-learn, which standardized the data to have a standard deviation of 1 and a mean of 0 (Testas, 2023). The dataset was split into training and test sets (80:20) to effectively evaluate the model's performance on unseen data. The classifier was then trained on the training dataset and benchmarked against the test set. The dataset was split into training (497 samples) and testing sets (125 samples), each with 11 features extracted from the molecular features of the LNP SMILE strings. Next, the RFC was deployed using the scaled training data, and the performance was evaluated on the test set. The feature importance was analyzed to identify the most influential factors for TE and evaluated using five-fold cross-validation. The molecular features (x) are separated from the target variable of transfection efficiency (y). The features were standardized using StandardScaler from scikit to ensure a mean of 0 and a standard deviation of 1. The RFC was initialized with 100 trees and a fixed random state for reproducibility. Five-fold cross-validation was then conducted using the cross_val_score function, in which the data was split into five subsets. The model was trained on four subsets and tested on the remaining one, repeating this process five times to ensure each subset was used as a test set once. The cross-validation scores were then calculated for each fold, providing an estimate of the model's performance, as shown in Fig. 3.3. Appendices B and C describe the detailed code and links for all models and datasets.

The RFC was evaluated using five-fold cross-validation with cross-validation scores of 0.824, 0.824, 0.782, 0.815, and 0.484. The mean accuracy was 0.746, with a standard deviation of accuracy of 0.132. This indicates that the classifier performed relatively well, considering the level of variability across the different folds. Still, the lower score in the fifth fold suggests that further model refinement is necessary or that data imbalance is a problem. Figure 3.4 below shows a Receiver Operating Characteristic (ROC) curve. The ROC is used to represent a model's performance, plotting the TPR against the FPR at different threshold settings (Gönen, 2006; Oh et al., 2021). The ROC curve shows that the model effectively separates the positive and negative classes, yielding an AUC of 0.92. This means the classifier effectively separates the two classes, as an AUC of 1 represents the threshold for a perfect model, and an AUC of 0.5 represents the threshold for a model performing no better than random chance (Carrington et al., 2021). There is a clear separation between the positive class and the negative class.

This high AUC score indicates that the model can distinguish between LNPs with high and low transfection efficiencies, correctly identifying positive cases (i.e., true positives) in 89 instances, while maintaining a strong balance by accurately recognizing 453 negative cases (i.e., true negatives). The ROC curve, which remains close to the top-left corner of the plot, demonstrates the model's ability to balance sensitivity, i.e., the true positive rate, and specificity, i.e., the true negative rate, highlighting relative reliability. It has a relatively low false positive rate, incorrectly labeling only 34 instances as positive, but it missed 46 positive instances, categorizing them as negative. These false negatives can be a concern where accurately

```
                                           combined_smiles  Transfection_Efficiency  \
0  O=P(OCCCC)([O-])OCC[NH2+]CCCC.CCCCCCCCCCCCCCC...                                0
1  O=P(OCCCCC)([O-])OCC[NH2+]CCCC.CCCCCCCCCCCCCC...                                0
2  O=P(OCCCCCC)([O-])OCC[NH2+]CCCC.CCCCCCCCCCCCC...                                0
3  O=P(OCCCCCCC)([O-])OCC[NH2+]CCCC.CCCCCCCCCCCC...                                0
4  O=P(OCCCCCCCC)([O-])OCC[NH2+]CCCC.CCCCCCCCCCC...                                0

   Molecular_Weight  Number_of_Rings    logP  Polar_Surface_Area  NumHDonors  \
0          3685.201                4  31.719              575.85           2
1          3699.228                4  32.109              575.85           2
2          3713.255                4  32.499              575.85           2
3          3727.282                4  32.889              575.85           2
4          3741.309                4  33.279              575.85           2

   NumHAcceptors  NumRotatableBonds  unsaturation  ionizable  helper_lipid  \
0             56                214             4          1             1
1             56                215             4          1             1
2             56                216             4          1             1
3             56                217             4          1             1
4             56                218             4          1             1

   pegylated
0          1
1          1
2          1
3          1
4          1
Cross-validation scores: [0.824      0.824      0.78225806 0.81451613 0.48387097]
Mean accuracy: 0.7457290322580644
Standard deviation of accuracy: 0.13182440564249429
```

Fig. 3.3 Initial RFC with five-fold cross-validation results

predicting TE is vital for predictive treatment efficacy. Evaluating the model's performance required a comprehensive approach, focusing on several metrics beyond accuracy. Precision, recall, and F1 scores were calculated to understand the model's performance across different aspects. Precision indicated the proportion of true positive predictions among all positive predictions, while recall measured the ability to identify all relevant instances. The F1 score balanced precision and recall to capture the model's effectiveness. These metrics were computed using the accuracy_score, precision_score, recall_score, and f1_score functions from scikit-learn, providing a detailed picture of the RFC's predictive capabilities.

The results from the five-fold cross-validation for the RFC yielded are shown in Table 3.2. The accuracy indicates that the initial RFC model correctly predicted TE about 87.13% of the time, with the low standard deviation suggesting that performance is consistent across different folds. The precision, i.e., the proportion of correct positive identifications, was 73.09%, indicating that the model predicts positive TE about 73.09% of the time (Powers, 2020). The higher standard deviation indicates variability in precision across different folds. The recall, i.e., the proportion of actual positives that were correctly identified, was 65.93%, indicating that the model identified approximately 65.93% of actual positive TEs. The high standard deviation is associated with variability in the model's ability to identify positive cases across different folds (Najdi, 2018). The F1 score, i.e., the harmonic mean of

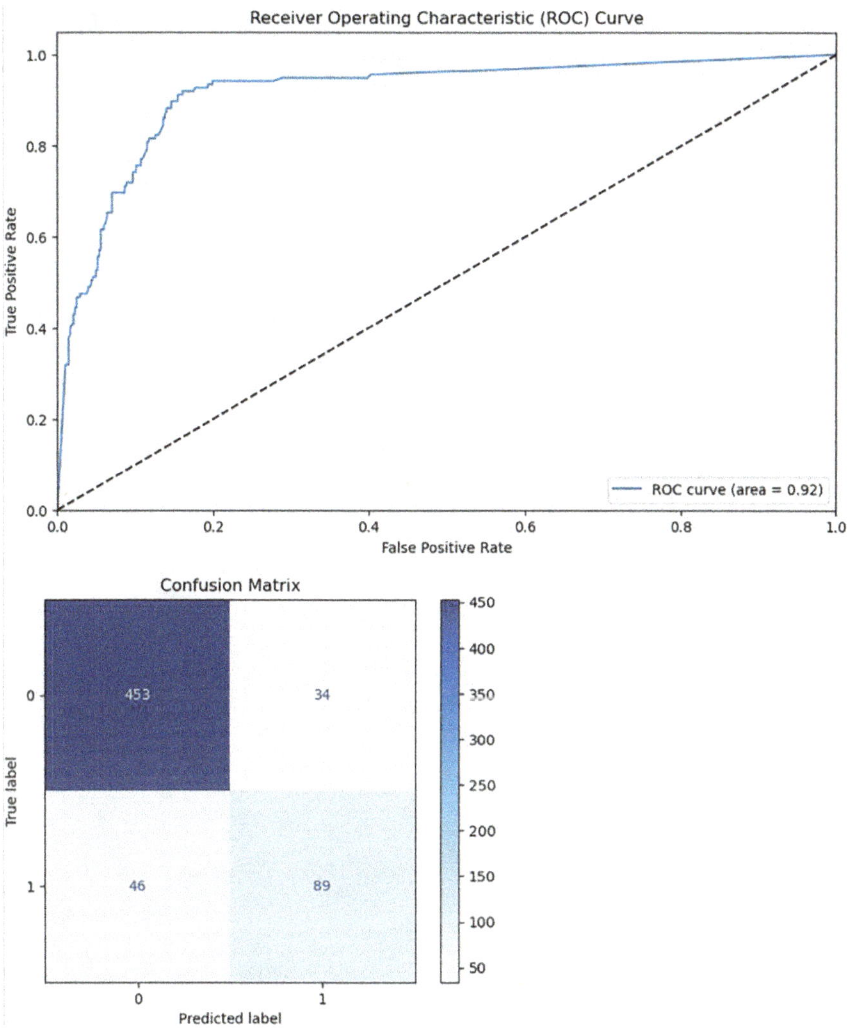

Fig. 3.4 ROC and Confusion Matrix for RFC

precision and recall, of 68.52%, means a good balance between precision and recall, with some variability across folds (Ali et al., 2023a, b). The RFC demonstrated good performance with an accuracy of approximately 87%. However, there is variability in precision, recall, and F1 scores across different folds, particularly in recall. The variability means that while the model is generally reliable, its ability to identify all positive cases, i.e., correctly recall, is not as consistent (Powers, 2020).

In developing the RFC model, class imbalance was observed in the dataset, which contained a disproportionate number of efficient versus inefficient transfections. It was evident that more advanced techniques and future improvements were needed to address this, such as applying SMOTE to generate synthetic samples for

Table 3.2 RFC
metric results

Metric	Value (Mean ± Std Dev)
Accuracy	0.8713 ± 0.0257
Precision	0.7309 ± 0.0744
Recall	0.6593 ± 0.1248
F1-score	0.6852 ± 0.0764
AUC-ROC	0.92

the minority class. Additionally, feature engineering posed challenges, as selecting the most relevant features required a balance between retaining important information and avoiding redundancy (Li et al., 2017). The insights gained from this initial model laid the groundwork for exploring more sophisticated techniques, such as SSL and advanced algorithms, including XGBoost. Each stage of the process, from data preparation to model evaluation, was an iterative learning experience that informed the development of more effective models, which will be discussed in the next chapter.

Chapter 4
Semi-Supervised Machine Learning Implementation & Results

With the baseline RFC model implemented for TE prediction, an SSL model with pseudo-labeling was explored. There are vast troves of biological data with insufficiently labeled, structured, and tagged data. Thus, in a real-world context, there is often only access to a small proportion of sufficiently structured and labeled data that can be utilized for experimentation and foundational model development. The RFC was trained using only the labeled data, and its performance was evaluated through five-fold cross-validation. The trained model was then used to predict labels for the unlabeled data, generating pseudo-labels. The pseudo-labeled data were combined with the original labeled dataset, creating an augmented training set that provided a more comprehensive representation of the data. The model was then retrained on this combined dataset, enabling it to learn from a broader range of examples and thereby improve its accuracy.

4.1 Pseudo-Labeling with Confidence Thresholding

As previously discussed, the initial RFC model achieved a mean accuracy of 0.746, with cross-validation scores ranging from 0.484 to 0.824, indicating variability and further model refinement. To address this, an SSL approach with iterative pseudo-labeling was developed. The RFC was trained on the labeled dataset. Then, the trained model was used to predict labels for the unlabeled data, and the most confident predictions with a confidence threshold of 0.9 were selected for inclusion in the training set (Ali, 2021). This process was repeated iteratively, gradually expanding the labeled dataset with high-confidence pseudo-labeled data. In addition, the importance of the molecular feature was reviewed by examining the features to see if the model is overly reliant on any particular features, which might indicate potential overfitting. A data split strategy was utilized to validate these results and ensure

K. W. Ramadurai, A. Banerjee, *Machine Learning-Driven Rational Design in Nanomedicine*, SpringerBriefs in Bioengineering, https://doi.org/10.1007/978-3-032-04012-1_4

no overlap between training and test sets. The model was retrained using the SSL approach on the training set and then evaluated on a separate test set that was not used in the training or pseudo-labeling process. Cross-validation with shuffling is then performed to ensure the model's performance is consistent across different data splits. The evaluation of the SSL model on the separate test set resulted in an accuracy of 0.861 (Fig. 4.1). The test set accuracy indicates that the iterative

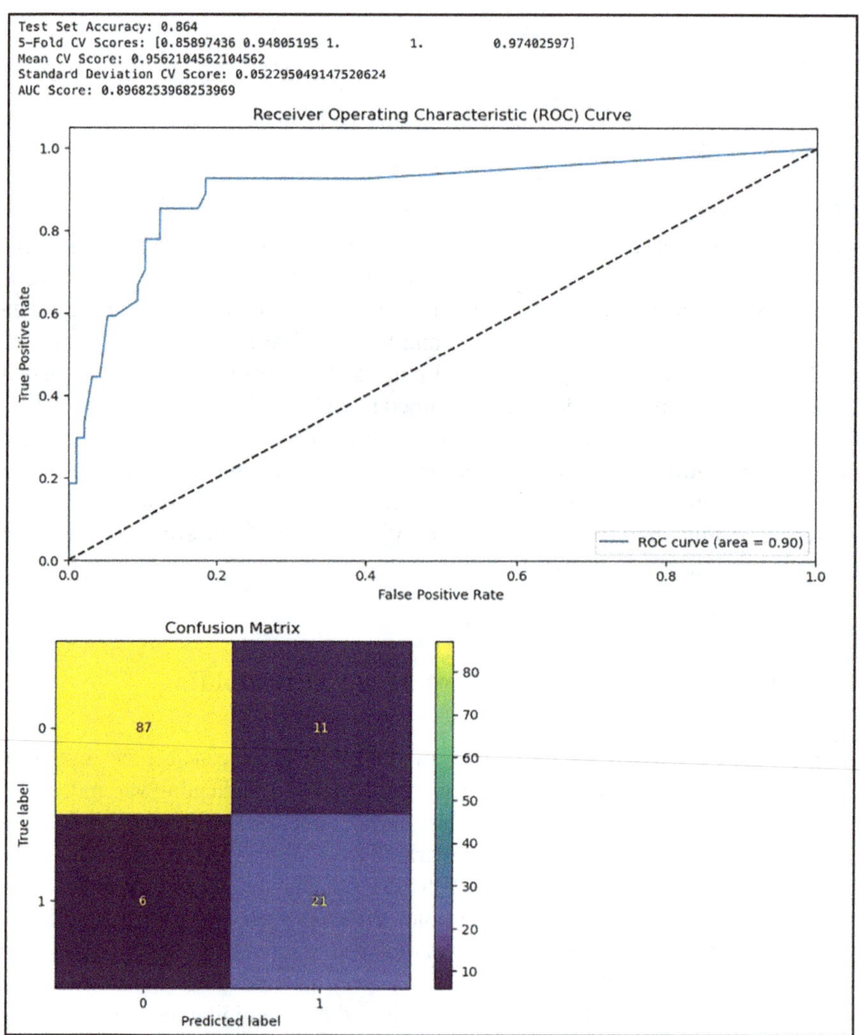

Fig. 4.1 SSL model five-Fold CV results, ROC curve, and confusion matrix

training process influenced the initial results and improved over a supervised approach. Feature importance analysis was conducted to ensure a balanced utilization of the dataset's features. Key features, including molecular weight, logP, and polar surface area, were identified as the most influential. This balanced feature importance distribution indicates that the model does not overfit any particular feature. Probabilistic modeling further validated the model, incorporating only high-confidence predictions into the training set to reduce variability and enhance accuracy. The iterative inclusion of confident pseudo-labels allowed the model to learn from a broader dataset, ultimately improving its predictive capabilities. The revised model test set accuracy of 0.861 confirms that the initial high accuracy observed during cross-validation was influenced by the iterative pseudo-labeling process, which gradually incorporated pseudo-labeled data. This realistic test accuracy, obtained from a separate test set, indicates that the revised model's performance is not due to data leakage but reflects a true generalization capability (Avrithis, 2021).

Once the new SSL model was run, a ROC curve was created. Figure 4.1 shows that the AUC of the model distinguishing between positive and negative classes is 0.90, and it also displays the confusion matrix for the model. It can be observed that the true positives align relatively well with the true negatives. There was a low number of false positives and false negatives. The model correctly predicted 87 instances as true negatives and 21 as true positives. There were 11 false positive mistakes, i.e., predicting a positive outcome when the actual outcome was negative. There were also six false negative errors, i.e., wrongly predicting the class to be negative when it was positive. Precision, recall, and F1 scores were calculated using the matrix above. An F1-score of 0.712 indicates that the ML model strikes a good balance between minimizing false positives and maximizing true positives. The value of 0.656 precision suggests that the ML model performs slightly better on true positives, while a recall value of 0.778 indicates that the ML model performs slightly less effectively in avoiding false positives. Overall, the SSL model with iterative pseudo-labeling performs well in predicting TE, as evidenced by its high AUC and good precision, recall, and F1 scores.

$$\textbf{Precision}\left(\text{Positive Predictive Value}\right): \text{Precision} = \frac{TP}{TP+FP} = \frac{21}{21+11} = 0.656$$

$$\textbf{Recall}\left(\text{Sensitivity}\right): \text{Recall} = \frac{TP}{TP+FN} = \frac{21}{21+6} = 0.778$$

$$\textbf{F1-Score}: \text{F1-Score} = 2\times\frac{\text{Precision}\times\text{Recall}}{\text{Precision}+\text{Recall}} = 2\times\frac{0.656\times0.778}{0.656+0.778} = 0.712$$

4.2   Synthetic Minority Over-Sampling Technique & XGBoost Model Enhancement

While these models demonstrate solid benchmark performance in TE prediction, additional tools for model refinement and optimization were explored. This includes hyperparameter tuning via GridSearchCV to find the optimal hyperparameters for the RFC and synthetic data generation utilizing the Synthetic Minority Over-sampling Technique (SMOTE) to address class imbalances within the dataset. Often, class imbalance hinders the performance of ML models, leading to biased predictions and suboptimal results (Dube & Verster, 2024). Applying SMOTE-generated synthetic samples for the minority class balances the dataset, enabling the model to learn more effectively from both classes (Joloudari et al., 2023; Wongvorachan et al., 2023). After balancing the dataset, this approach was integrated into the SSL framework with iterative pseudo-labeling. This combined approach facilitated the iterative enhancement of the labeled dataset with high-confidence pseudo-labeled data drawn from a more balanced and representative sample space. The refined model, trained with the optimized hyperparameters identified through GridSearchCV, demonstrated significant performance improvements. Specifically, the SMOTE-enhanced SSL model achieved a higher accuracy on the test set and improved ROC AUC scores, indicating its enhanced ability to distinguish between different TEs, with strong generalization capabilities and optimized performance (Fig. 4.2). The balanced dataset ensured that the model was accurate and generalizable, thereby reducing the risk of overfitting to a specific class (Wongvorachan et al., 2023). This shows the importance of addressing class imbalance and the efficacy of combining data augmentation techniques with ML frameworks to optimize predictive models for LNP TE.

Using the confusion matrix, the precision, recall, and F1 scores were calculated as follows. The model had a precision of 0.657, indicating that it makes some false positive predictions, which is to be expected in a balanced/imbalanced dataset where synthetic samples are introduced. A recall of 85.2 means that the model identifies a high proportion of true positive cases, critical in therapeutic development applications where missing positive cases can have severe consequences for patients. The F1 score of 0.742 strikes a good balance between precision and recall, and an AUC score of 0.93 indicates that the model can effectively distinguish between positive and negative classes.

Precision (Positive Predictive Value):

$$\text{Precision} = \frac{TP}{TP + FP} = \frac{23}{23 + 12} \approx 0.657$$

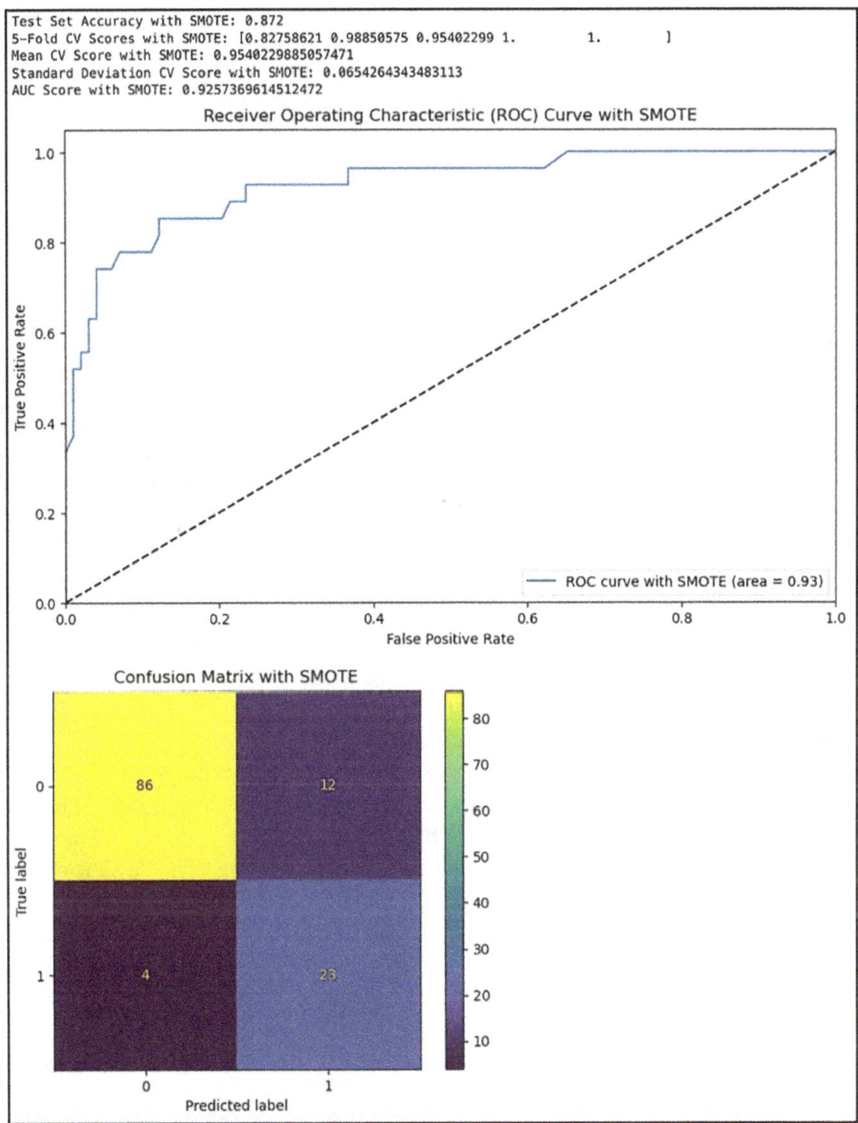

Fig. 4.2 SMOTE model five-Fold CV Results, ROC curve, and Confusion Matrix

Recall$($Sensitivity or True Positive Rate$)$:

$$\text{Recall} = \frac{TP}{TP + FN} = \frac{23}{23 + 4} \approx 0.852$$

F1 - Score :

$$\text{F1 - Score} = 2 \times \frac{\text{Precision} \times \text{Recall}}{\text{Precision} + \text{Recall}} = 2 \times \frac{0.657 \times 0.852}{0.657 + 0.852} \approx 0.742$$

The results show that integrating SSL models with iterative pseudo-labeling, hyperparameter tuning, and synthetic data generation enhances the prediction of TE. The SMOTE-enhanced model demonstrated higher accuracy and generalization capabilities, effectively addressing class imbalance and optimizing overall accuracy. These methods aim to further refine and enhance in silico virtual screening, addressing key challenges such as class imbalance and parameter optimization, thereby contributing to the development of more accurate and reliable predictive models for LNP TE. The models developed can be applied to unstructured and unlabeled LNP data by following a systematic preprocessing, feature extraction, and iterative pseudo-labeling framework. Using techniques, including SMOTE, to address the class imbalance, combined with hyperparameter tuning, ensures that the models are accurate and generalizable (Joloudari et al., 2023).

Additional research was conducted to optimize the model further using the SMOTE-enhanced dataset and XGBoost. XGBoost significantly improves the model by utilizing advanced gradient-boosting techniques to enhance prediction accuracy (Ali et al., 2023a, b). It works by sequentially building a series of decision trees, where each tree corrects errors made by the previous ones (Ali et al., 2023a, b). This process minimizes bias and variance in the dataset, incorporates regularization to prevent overfitting, handles missing data efficiently, and supports parallel processing, making it faster and more scalable (Chen & Guestrin, 2016; Quinto, 2020). In this case, XGBoost's ability to handle complex data patterns and improve classification precision and recall made it helpful in predicting LNP TE. The results are presented in Fig. 4.3, with the comparative analysis of models shown in Fig. 4.4. The model achieved an accuracy of 0.9314, indicating that 93.14% of the instances the model predicted as positive were correct. This corresponds to a recall of 0.9794, meaning that for the actual positive instances, the model correctly identified 97.94% of them. The model also achieved an F1 Score of 0.9548. The model achieved an accuracy of 95.38% and an AUC of 0.9862, signifying good overall performance and class ability. Additionally, the figure of 93.14% precision and 97.94% recall demonstrates both precision with very few false positives and sensitivity with very few false negatives. The confusion matrix shows an accurate classification of the model, with 91 out of 98 negative cases and 95 out of 97 positive cases correctly classified. To illustrate cross-validation consistency, the best cross-validation score was 0.9038, with a mean of 0.7945, indicating a good model that performs well across different data splits. The test set accuracy of 0.9538 also confirms that it generalizes well to unseen data. Overall, the high precision, recall, and F1 score indicate that the XGBoost model is an effective predictive model for in silico screening applications, with minimal false positives and false negatives. While the results demonstrate the potential of these ML models to predict LNP TE

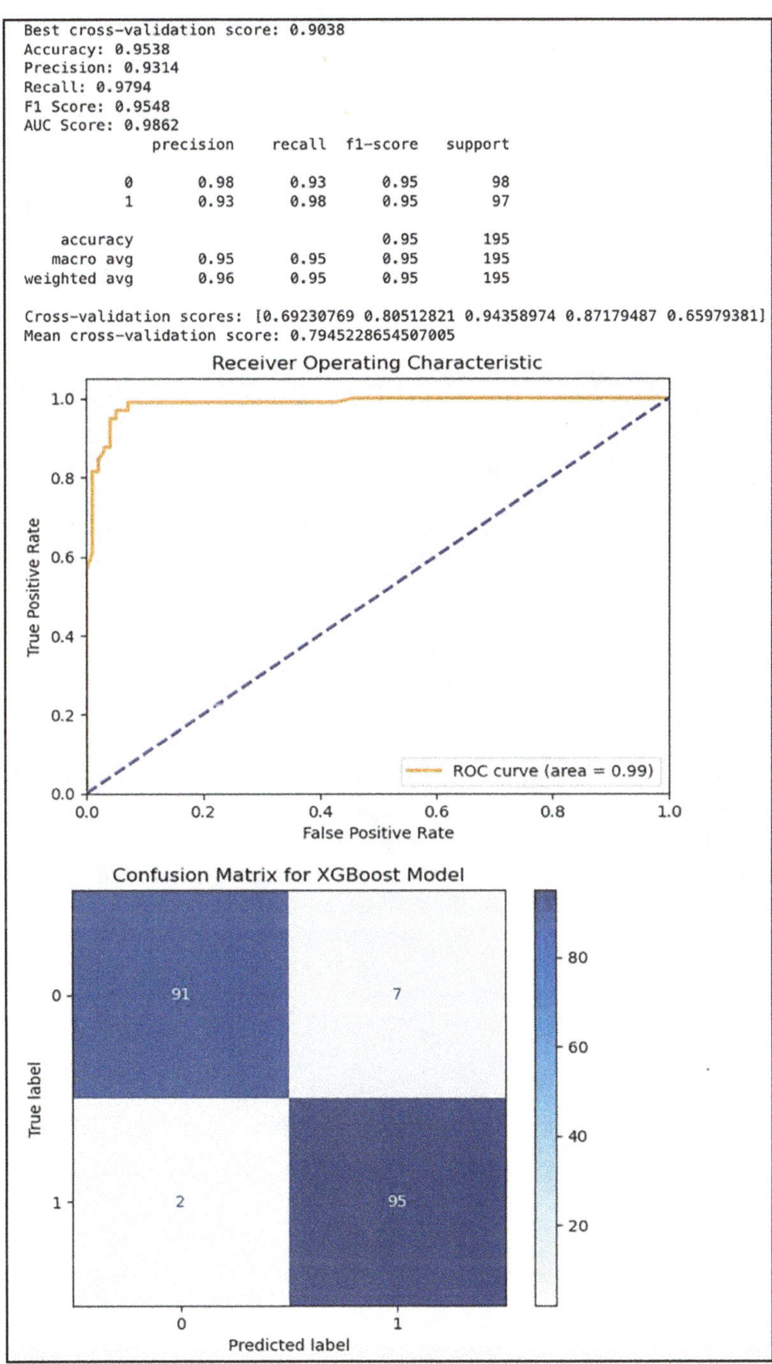

Fig. 4.3 Performance of the XGBoost model on the SMOTE-augmented dataset

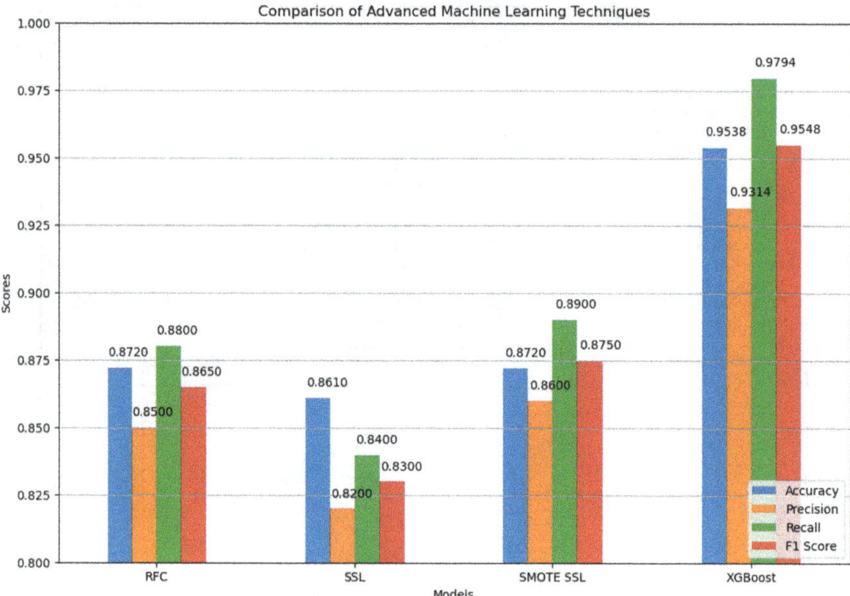

Fig. 4.4 Comparative analysis of ML techniques

accurately, it is essential to take a macro perspective and assess the broader implications for developing and integrating next-generation computational modalities in
biopharmaceutical research and development. In the next section, we discuss how
these results can be applied to practical applications within nanomedicine, outline
the research challenges and limitations of the models, and propose future directions for research.

Chapter 5
Discussion of Modeling Techniques, Practical Implications, and Prospective Developments

This research explored the functional integration of cheminformatics approaches and ML-enhanced virtual in silico screening techniques to improve LNP TE prediction. The XGBoost model significantly enhanced the performance of the SMOTE SSL model, achieving a test set accuracy of 95.38%, a precision of 93.14%, a recall of 97.94%, and an F1 score of 95.48%. These metrics indicate an effective model with minimal false positives and false negatives, significantly enhancing the reliability of TE predictions. The high AUC score of 0.9862 confirms the ability to distinguish between positive and negative transfection outcomes, which has substantial implications for the development and optimization of nanocarriers for nanomedicines. High TE is critical for ensuring that mRNA is effectively delivered and expressed within target cells, in which accurate predictions of TE can streamline therapeutic design and testing phases, reducing time and costs (Meyer et al., 2022; Ouranidis et al., 2021; Wang et al., 2024).

Additionally, the ability to handle class imbalances and leverage both labeled and unlabeled data improves the generalizability of the models, making them more applicable to real-world scenarios where labeled data may be scarce. This advancement supports the broader objective of creating more effective and safer mRNA-based therapies within the clinical domain of oncology. The SMOTE-enhanced model validates the effectiveness of utilizing SMOTE to address class imbalance challenges and leverage unlabeled data to address data sparsity challenges that often complicate modeling in biotechnology research. One point of particular interest is the model's ability to maintain high recall while improving overall accuracy. This suggests it is highly sensitive in identifying true positive cases, which is critical when selecting suitable nanocarriers for therapeutic development. These findings demonstrate that combining SSL with advanced oversampling techniques, such as SMOTE, can yield better results. The iterative pseudo-labeling process enabled continuous learning and model refinement, demonstrating high accuracy in predictions and suggesting that it can reliably

K. W. Ramadurai, A. Banerjee, *Machine Learning-Driven Rational Design in
Nanomedicine*, SpringerBriefs in Bioengineering,
https://doi.org/10.1007/978-3-032-04012-1_5

differentiate between high and low transfection efficiency LNPs. Furthermore, feature importance insights were garnered to understand which molecular features most influence TE and guide the rapid screening and design of new LNPs with optimized nanomedicine properties.

The results from the baseline RFC and the subsequent enhancements using SSL, SMOTE, and XGBoost demonstrate a progressive improvement in predicting LNP TE. The initial implementation of the RFC provided a solid starting point, achieving an accuracy of 74.6% with a cross-validation accuracy mean of 0.746 and variability across folds. The AUC of 0.92 indicates that the model can effectively distinguish between high- and low-TE LNPs. However, the model's recall (65.93%) demonstrated a moderate capability to identify all relevant instances, particularly in detecting positive TE cases. This variability necessitated further refinement, as accurately identifying positive cases is crucial to ensuring the proper development of lead therapeutic candidates. The SSL model improved, achieving a test set accuracy of 86.1%, an increase from the baseline. The AUC-ROC of 0.90 indicated class distinguishability (Table 5.1). The precision (65.6%) and recall (77.8%) demonstrated a balanced performance, with it being slightly more effective at identifying true positives than minimizing false positives (Table 5.1). This balance is reflected in the F1 score of 71.2%, indicating that the SSL approach successfully leveraged unlabeled data to enhance accuracy.

Given the class imbalance in the dataset, the Synthetic Minority Over-sampling Technique (SMOTE) was applied to generate synthetic samples for the minority class, thereby balancing the dataset. The accuracy remained at 86.1%, consistent with the SSL model, but the recall improved to 85.2%, thereby enhancing the ability to identify true positive cases correctly. The precision of 65.7% and an F1 score of 74.2% further suggest that the model achieved a good balance between precision and recall. The AUC-ROC score of 0.93 reinforced the ability to distinguish between different TEs effectively. This balanced performance highlights the importance of addressing class imbalance, especially concerning in silico screening, where there are potentially thousands of LNPs that likely have insufficient TE.

To further enhance the performance, XGBoost was employed, which significantly improved the results. The XGBoost model achieved an accuracy of 95.38%, far surpassing the previous models. The AUC-ROC score of 0.9862 indicated an ability to differentiate between the classes, close to the ideal score of 1. The precision and recall rates were particularly noteworthy, at 93.14% and 97.94%, respectively, indicating high precision with few false positives and high sensitivity with few false negatives. The F1 score of 95.48% reflected a balance between precision

Table 5.1 Key performance metric summary

Model	Accuracy	Precision	Recall	F1-Score	AUC-ROC
Baseline RFC	0.746	0.7309 ± 0.0744	0.6593 ± 0.1248	0.6852 ± 0.0764	0.92
SSL	0.861	0.656	0.778	0.712	0.90
SMOTE	0.861	0.657	0.852	0.742	0.93
XGBoost	0.9538	0.9314	0.9794	0.9548	0.9862

and recall, suggesting that XGBoost and advanced gradient-boosting techniques are highly effective in handling complex data patterns and optimizing predictive accuracy in LNP TE. The progression from the baseline RFC to the XGBoost model illustrates a clear optimization trend with iterative model refinement.

5.1 Impact and Ethical Considerations

The models explored and implemented in this book demonstrate the power of ML-enhanced in silico virtual screening to accelerate preclinical drug development, focusing on LNPs that are likely to yield the best translatability and reduce trial-and-error experimentation. This has broader implications for the pharmaceutical industry, where rapid and cost-effective development of new therapies is critical. Specifically, in the context of mRNA vaccines, these models could accelerate the design of more effective nanocarriers, thereby improving therapeutic outcomes and paving the way for more personalized treatment approaches. The ability to efficiently optimize LNP formulations could also support the broader application of mRNA-based therapies beyond oncology, extending to areas including infectious diseases, neurodegenerative conditions, and autoimmune diseases.

Employing both supervised and semi-supervised learning models, along with parameter optimization, SMOTE, and XGBoost, demonstrated the potential to enhance predictive accuracy and reliability in identifying effective LNP formulations. The application of supervised models provided a good baseline with high accuracy, and utilizing SMOTE to handle class imbalance further refined these models, ensuring a better balance between precision and recall. This enhancement is critical as it ensures that fewer effective LNP formulations are overlooked, which is vital for successfully delivering mRNA vaccines. Due to the scarcity of labeled data in this domain, a semi-supervised learning scenario was simulated by splitting a single labeled dataset into a small labeled subset and a larger unlabeled subset. This approach enabled the leveraging of available labeled data to generate additional training data, thereby improving the model's performance. The SSL approach significantly improved the model's ability to generalize from limited labeled data by iteratively pseudo-labeling and expanding the labeled dataset. This is particularly relevant in real-world scenarios where obtaining labeled data is expensive and time-consuming. SSL thus offers a practical solution to maximize the utility of available data (Kim & Shin, 2013). Incorporating molecular feature/structural analysis via molecular fingerprinting data allowed for a deeper understanding of the structural properties influencing LNP efficacy. This analytical layer improved the model's predictive capabilities and provided insights into the critical molecular characteristics that drive successful transfection. In addition, this research demonstrated the unique ability to create an integrative tech stack of layered analysis and application frameworks that span chemoinformatic and ML domains. This shows the effectiveness of an integrative approach to therapeutic development, which combines computer science, structural biology, and medicinal chemistry domains to accelerate research and development, ultimately leading to the creation of novel nanomedicines.

The models explored can reduce the time and cost associated with preclinical drug development (Honkala et al., 2022; Kiriiri et al., 2020). By accurately predicting the TE of various LNP formulations, researchers can rapidly identify the most promising candidates for further development and more strategically allocate resources. The functional integration of these virtual screening in silico models into pharmaceutical therapeutic development pipelines will inevitably change biotechnology workflows. It is well known that it often takes over a decade and hundreds of millions, if not billions, to bring a therapy to and through clinical trials to FDA or EMA approval. Integrating next-generation in silico virtual screening models could significantly accelerate the front-end preclinical drug development process, ultimately reducing the capital costs associated with therapeutic asset development. Beyond mRNA vaccines, the principles of molecular feature analysis and advanced ML-enabled virtual screening can be applied to predict the efficiency of numerous nanoparticle-based delivery systems, including those used in gene therapy. This makes the technology versatile and broadly applicable across different therapeutic domains with high unmet patient needs. Ideally, these compute-enabled in silico screening models would be functionally integrated into a full-stack, end-to-end preclinical research workflow that can eliminate manual processes and experimental redundancies.

These models serve as the foundation for digitizing the entire preclinical drug development workflow, encompassing drug discovery, design, screening, and in vitro and in vivo testing. The integration with next-generation approaches, including digital twinning, which promises to create entire digital models of biological organisms such as transgenic mice, can ultimately eliminate the need for redundant animal models, which often do not translate well to humans and clinical subjects. The horizon for these next-generation technologies has been accelerated with the recent advancements in computing technologies, including deep learning and deep neural networks, which could potentially unlock the ability to model complex biological systems in an ex vivo format digitally. This means that researchers can model the dynamic interaction of therapeutic compounds outside the body before administering them to an animal or human. The ability to run advanced biosimulation and digital twin modeling of complex biological organisms has only recently become viable with the advent of deep neural networks, which can execute highly parallelized simulations and integrate the hundreds of millions of data points required for sufficient simulations of complex physiological systems.

While the advancements in in silico screening and nanomedicine design offer significant benefits, they also raise several ethical considerations. When developing personalized therapeutics that utilize patient data and real-world evidence (RWE), ensuring patient data privacy, compliance, and security is paramount, particularly with sensitive health information. This includes implementing data anonymization techniques and access controls to protect patient confidentiality (Pillai et al., 2022; Zou et al., 2022). Secondly, the potential for biases in the data and models must be addressed to avoid disparities in treatment outcomes, as models trained on biased data can perpetuate existing inequalities (Moore, 2022). It remains crucial to utilize diverse, heterogeneous, and representative datasets and to regularly audit models for fairness and bias (Moore, 2022). Additionally, transparent reporting of model performance and limitations is crucial for maintaining transparency in the development of these technologies. Thus, researchers must communicate their models' strengths, weaknesses, and biases, ensuring that stakeholders understand the potential risks and benefits to define their relative applications and limitations. The open-source access to these models, coupled with information transparency, is vital in building highly diverse and heterogeneous foundational models that can be iteratively refined as technologies and data continue to expand and evolve.

5.2 Limitations and Challenges

While the models explored in this book demonstrate strong screening capabilities, several limitations should be acknowledged. One key limitation is the generalizability of the models to new datasets that differ significantly from the training data. For instance, LNP formulations not included in the original dataset may exhibit characteristics that the model has not been trained to recognize, potentially leading to reduced accuracy. This highlights the need for ongoing model validation using diverse datasets to ensure the applicability across different LNP designs and therapeutic applications. Another limitation is the assumption of uniformity in the experimental conditions across all data points. Variations in experimental setups, such as differences in how TE is measured or discrepancies in LNP formulations, could introduce biases that the model may not account for. Additionally, while SMOTE addressed class imbalance, generating synthetic samples may introduce noise or distort the underlying data distribution, potentially affecting the model's performance in real-world scenarios. Future research should explore alternative methods for handling imbalances and investigate how the model's performance holds up when applied to clinical or preclinical experimental data that vary from the initial dataset.

The pseudo-labeling approach relies on the accuracy of the initial model's predictions. If the initial model is not sufficiently accurate, the pseudo-labels may introduce noise into the training data, potentially degrading the final model's performance. Other limitations include potential biases in the dataset and the model's generalizability. While SMOTE was utilized to balance the classes, subtle biases could be present if the synthetic samples do not represent the real data distribution. Also, the size and diversity of the dataset used in this study were limited to the only publicly available dataset on LNP transfection efficiency. It is important to note that this research was framed as a functional exploration of supervised and semi-supervised learning techniques and their biological applications. Due to the scarcity of labeled data available, a functional simulation of a semi-supervised learning approach was developed by splitting the single labeled dataset into a small labeled subset and a larger unlabeled subset. This approach enabled the leveraging of available labeled data to generate additional training data, thereby improving the model's performance. However, this method introduces limitations and potential biases, such as the risk of overfitting to the initially labeled subset and the propagation of incorrect pseudo-labels during training (Arazo et al., 2020; Li et al., 2023). Additionally, the representativeness of the split datasets is crucial; if the labeled subset is not representative, the model may not generalize well (Li et al., 2023). Despite these limitations, this approach provides a pragmatic solution to maximize available data, advancing the understanding and predictive capabilities in the context of transfection efficiency for nanomedicines.

Once again, although the dataset provided valuable insights, a more extensive and diverse dataset would likely yield better results, as it does not fully capture the full diversity of real-world LNP formulations and their transfection efficiencies, which could lead to biases if the model encounters unseen variations in the new datasets. The simulated model's performance would degrade if the new datasets had significantly different features or distributions. Additionally, only a handful of chemoinformatic molecular features were crucial in the model development. Therefore, a more comprehensive analysis is necessary to investigate additional chemoinformatic features and structural motifs that can be leveraged to enhance the in silico screening process. Additionally, the complexity of LNP formulations and the inherent biological variability in transfection efficiencies pose challenges that must be addressed (Guasp et al., 2024; Mendes et al., 2022). Future studies should validate these findings using larger datasets and explore the integration of other data sources, such as genomic and proteomic data (Guasp et al., 2024; Mendes et al., 2022; Sork, 2018). Further understanding the biological mechanisms underlying LNP transfection could lead to the development of more biologically informed models. TE is only one functional aspect in a host of critical molecular and biological interactions, including endosomal escape. This research only covered a handful of molecular features, given time and resource constraints, and numerous elements beyond just structural motifs influence nano-oncological development. While relying on synthetic data generated by SMOTE is beneficial for addressing class imbalance, it may introduce noise and affect model generalization. Additionally, the model's performance may vary with different datasets, highlighting the need for further validation across diverse data sources. This research was limited to a small scope of transfection efficiency prediction, as numerous other molecular and chemoinformatic factors influence LNP efficacy for therapeutic applications. The molecular features were also limited to only those that could be adequately and correctly quantified utilizing the available molecular and chemoinformatic tools. Notably, these models demonstrate a proof-of-concept application of integrating numerous application tools to accelerate the functional high-throughput screening of molecular compounds, specifically LNPs, that are best suited for therapeutic applications. There was also a functional limitation on the number of molecular features that could be suitably extracted from the SMILES strings, in which critical features such as zeta potential could not be reliably calculated due to the need for advanced chemoinformatic programs that were only accessible to industry professionals. Ideally, a larger host of molecular and chemoinformatic features would have been calculated and extracted to provide a more comprehensive picture.

5.3   **Future Research Directions**

One of the most promising directions for future research is the integration of synthetic data augmentation techniques. Synthetic data augmentation involves generating additional training data that mimics the properties of the original dataset (Jordon et al., 2022). Techniques such as variational autoencoders (VAEs) can be applied to create realistic synthetic data that enhances the diversity and size of the training dataset (De Silva & Brown, 2023; Girin et al., 2020; Joy et al., 2020). VAEs are a generative model that can improve the ML model by generating more realistic synthetic data and capturing intricate patterns. The VAE can be trained on the LNP TE data to learn its underlying distribution and develop new, realistic synthetic samples by sampling from the learned latent space and decoding these samples (Girin et al., 2020). This can augment the labeled dataset, which combines the synthetic data generated by the VAE with the labeled data and high-confidence pseudo-labeled data to create a diverse training set. Furthermore, combining SMOTE with other techniques, such as Tomek Links or Edited Nearest Neighbors (ENN), can remove noisy or redundant synthetic samples, thereby improving the quality of synthetic samples generated by SMOTE (Piotr & Rongbing, 2023).

In addition, continued refinement and molecular feature engineering can be conducted to create new features that might capture more complex relationships in the data. The implementation of hyperparameter tuning using methods such as Bayesian optimization could yield better performance, and integrating other unique data sources, including genomic, proteomic, and relevant clinical trial data, could enhance the model's applicability (Karadayı Ataş, 2024; Ujwal, 2021). As these tools develop, integrating improved and diverse heterogeneous datasets promises to refine and optimize nanocarrier modeling in complex biological systems. One of the most exciting directions is the implementation of real-time adaptive learning, whereby models can continuously improve as new data becomes available to ensure that these highly trained models remain relevant and accurate as our knowledge of biological interactions and biomaterials continues to expand (Kudithipudi et al., 2022). Creating an end-to-end in silico screening feedback loop and library, whereby new structural and molecular data of LNPs are fed into the model to optimize the next set of formulations, can further accelerate the design iteration of next-generation nanomedicines. This optimization would ideally yield therapeutic assets that are more translatable into viable clinical candidates. The ability to transition from simple in silico screening to the dynamic modeling of complex biological systems and organisms in an ex vivo context could represent a paradigm shift in how researchers approach therapeutic development in a systematically derisked manner. This can significantly alter the risk profile of biotechnology, which is often perceived as having a binary risk profile by the public and investors. Creating a unified, end-to-end, compute-enabled workflow that can digitalize and virtually model the entire preclinical development process could revolutionize biopharmaceutical development and usher in a new era of biotechnology, where biology finally becomes engineerable. This approach is finally on the cusp of realization with a new breed of biotechnology companies built from the ground up on AI-native computational platforms. These companies, such as Insilico Medicine, are integrating AI-enabled biosimulation modeling and drug design that can dramatically reduce the time and cost of preclinical drug development. These companies have broken the traditional model, which typically spans several years and tens of millions of dollars to bring a drug to the clinic, by achieving this goal in less than 18 months and under $2 million, representing a fundamental breakthrough in therapeutic research and development (Alcimed, 2021). The integration of heterogeneous patient-responder data will also be crucial in unlocking the full potential of these technologies and enhancing clinical translatability, as the therapeutic assets developed can be tailored to patient cohorts that are likely to be enrolled in trials, thereby overcoming a significant hurdle in translatability for therapeutics. Capturing and training these models on patient data can create de novo designer assets that can ideally perform well in stratified patient cohorts, thus reducing the high attrition rates in Phase I–III human clinical trials.

5.4 Conclusion

ML-enhanced virtual screening represents a powerful tool for optimizing LNPs for mRNA nanomedicines, setting the foundation for the clinical implementation of full-stack digital workflows in preclinical drug development. By leveraging data-driven insights and predictive modeling, researchers can streamline the discovery and development of new formulations, driving the future of personalized medicine and advanced therapeutics forward. The functional integration of advanced ML techniques with chemoinformatics, including molecular fingerprinting, pseudo-labeling, and synthetic data augmentation, can enhance the predictive accuracy and in silico screening of LNP TE using data-driven insights instead of trial-and-error experimentation. This proof-of-concept research demonstrates the feasibility of using advanced ML modeling combined with molecular fingerprinting to integrate the chemical structure of multicomponent LNPs to predict their functionality and provide a pragmatic solution to the limited availability of structured biological data. Iterative refinement and improvement of these types of functional models can accelerate the identification of optimal LNP formulations, leveraging molecular feature extraction. The enhanced model can then help select the most promising LNP candidates for experimental validation and streamline preclinical workflows. With the rapid emergence of next-generation computing modalities, including artificial intelligence, machine learning, and deep neural networks that can enable complex virtual biosimulation modeling, a new era of programmable and engineerable biology has come to fruition. These advanced digitalized approaches could dramatically alter the paradigm of taking decades of research and hundreds of millions of dollars to bring a regulatory-approved therapeutic market. These approaches would ideally lead to a significant reduction in experimental redundancies and costs, accelerate the discovery of insights, and ultimately democratize access to life-saving treatments at scale for millions of patients.

Appendix A Data Preparation and Processing

- Code snippet for concatenating SMILES strings and computing molecular feature descriptors

```python
import pandas as pd
from rdkit import Chem
from rdkit.Chem import Descriptors, Lipinski

file_path = 'Original_transfection_data.csv'
lnp_data = pd.read_csv(file_path)

# Compute molecular descriptors function
def compute_descriptors(smiles):
    mol = Chem.MolFromSmiles(smiles)
    if mol is None:
        return None
    descriptors = {
        'Molecular_Weight': Descriptors.MolWt(mol),
        'Number_of_Rings': Lipinski.RingCount(mol),
        'logP': Descriptors.MolLogP(mol),
        'Polar_Surface_Area': Descriptors.TPSA(mol),
        'NumHDonors': Lipinski.NumHDonors(mol),
        'NumHAcceptors': Lipinski.NumHAcceptors(mol),
        'NumRotatableBonds': Lipinski.NumRotatableBonds(mol),
        'unsaturation': sum(1 for bond in mol.GetBonds() if bond.GetBondType() in [Chem.rdchem.BondType.DOUBLE, Chem.rdchem.BondType.TRIPLE]),
        'ionizable': 1 if any(atom.GetFormalCharge() != 0 for atom in mol.GetAtoms()) else 0,
        'helper_lipid': 1,  # Binary
        'pegylated': 1  # Binary
    }
    return descriptors

# Combine SMILES strings into single representation
lnp_data['combined_smiles'] = lnp_data[['m1', 'm2', 'm3', 'm4']].apply(lambda x: '.'.join(x), axis=1)

# Compute descriptors for the combined SMILES string
descriptors = lnp_data['combined_smiles'].apply(lambda x: compute_descriptors(x))
descriptors_df = pd.DataFrame(list(descriptors))

# Concatenate the descriptors with the original dataset
lnp_data_with_descriptors = pd.concat([lnp_data, descriptors_df], axis=1)

# Drop rows with None values in descriptors
lnp_data_with_descriptors = lnp_data_with_descriptors.dropna()

# New dataset with calculated features
output_file_path = 'lnp_data_with_combined_features.csv'
lnp_data_with_descriptors.to_csv(output_file_path, index=False)

print(f"New dataset with calculated features saved to {output_file_path}")

New dataset with calculated features saved to lnp_data_with_combined_features.csv
```


K. W. Ramadurai, A. Banerjee, *Machine Learning-Driven Rational Design in
Nanomedicine*, SpringerBriefs in Bioengineering,
https://doi.org/10.1007/978-3-032-04012-1

- ## Code Snippet for Data Cleaning, Normalization, and Feature Extraction

```python
import pandas as pd

file_path = 'Finalized_ConcatenatedDissertation_LNP Data.csv'
data = pd.read_csv(file_path)

# Display the initial structure of the dataset
print(data.info())

# Handle missing values: Option 1 - Drop rows with missing values
data_cleaned = data.dropna()

# Handle missing values: Option 2 - Fill missing values (example: filling with mean or median)
# data_cleaned = data.fillna(data.mean())

# Encode target variable 'Transfection_Efficiency'
data_cleaned['Transfection_Efficiency'] = data_cleaned['Transfection_Efficiency'].astype('category').cat.codes

# Remove duplicates
data_cleaned = data_cleaned.drop_duplicates()

# Verify cleaned dataset
print(data_cleaned.info())


<class 'pandas.core.frame.DataFrame'>
RangeIndex: 622 entries, 0 to 621
Data columns (total 13 columns):
 #   Column                   Non-Null Count  Dtype
---  ------                   --------------  -----
 0   combined_smiles          622 non-null    object
 1   Transfection_Efficiency  622 non-null    int64
 2   Molecular_Weight         622 non-null    float64
 3   Number_of_Rings          622 non-null    int64
 4   logP                     622 non-null    float64
 5   Polar_Surface_Area       622 non-null    float64
 6   NumHDonors               622 non-null    int64
 7   NumHAcceptors            622 non-null    int64
 8   NumRotatableBonds        622 non-null    int64
 9   unsaturation             622 non-null    int64
 10  ionizable                622 non-null    int64
 11  helper_lipid             622 non-null    int64
 12  pegylated                622 non-null    int64
dtypes: float64(3), int64(9), object(1)
memory usage: 63.3+ KB
None
<class 'pandas.core.frame.DataFrame'>
Index: 588 entries, 0 to 621
Data columns (total 13 columns):
 #   Column                   Non-Null Count  Dtype
---  ------                   --------------  -----
 0   combined_smiles          588 non-null    object
 1   Transfection_Efficiency  588 non-null    int8
 2   Molecular_Weight         588 non-null    float64
 3   Number_of_Rings          588 non-null    int64
 4   logP                     588 non-null    float64
 5   Polar_Surface_Area       588 non-null    float64
 6   NumHDonors               588 non-null    int64
 7   NumHAcceptors            588 non-null    int64
 8   NumRotatableBonds        588 non-null    int64
 9   unsaturation             588 non-null    int64
 10  ionizable                588 non-null    int64
 11  helper_lipid             588 non-null    int64
 12  pegylated                588 non-null    int64
dtypes: float64(3), int64(8), int8(1), object(1)
memory usage: 60.3+ KB
None
```

```python
from sklearn.preprocessing import StandardScaler

# Extract features and target variable
X = data_cleaned.drop(columns=['Transfection_Efficiency', 'combined_smiles'])
y = data_cleaned['Transfection_Efficiency']

# Initialize the scaler
scaler = StandardScaler()

# Fit and transform the features
X_scaled = scaler.fit_transform(X)

# Convert the scaled features back into a DataFrame
X_scaled_df = pd.DataFrame(X_scaled, columns=X.columns)

# Combine the scaled features with the target variable
data_final = pd.concat([X_scaled_df, y.reset_index(drop=True)], axis=1)

# Display the final preprocessed data
print(data_final.head())
```

```
   Molecular_Weight  Number_of_Rings       logP  Polar_Surface_Area  \
0         -1.487486        -0.320771  -1.069914           -0.939749
1         -1.449112        -0.320771  -0.999898           -0.939749
2         -1.410738        -0.320771  -0.929882           -0.939749
3         -1.372363        -0.320771  -0.859866           -0.939749
4         -1.333989        -0.320771  -0.789851           -0.939749

   NumHDonors  NumHAcceptors  NumRotatableBonds  unsaturation  ionizable  \
0   -0.873974      -1.047676          -1.579042     -0.629448   0.167248
1   -0.873974      -1.047676          -1.529081     -0.629448   0.167248
2   -0.873974      -1.047676          -1.479120     -0.629448   0.167248
3   -0.873974      -1.047676          -1.429159     -0.629448   0.167248
4   -0.873974      -1.047676          -1.379198     -0.629448   0.167248

   helper_lipid  pegylated  Transfection_Efficiency
0           0.0        0.0                        0
1           0.0        0.0                        0
2           0.0        0.0                        0
3           0.0        0.0                        0
4           0.0        0.0                        0
```

- ## Molecular Feature Extraction Code Snippet

```python
from rdkit import Chem
from rdkit.Chem import Descriptors, MACCSkeys

# Function to calculate molecular descriptors
def calculate_descriptors(smiles):
    mol = Chem.MolFromSmiles(smiles)
    if mol:
        descriptors = {
            'MolWt': Descriptors.MolWt(mol),
            'LogP': Descriptors.MolLogP(mol),
            'NumHAcceptors': Descriptors.NumHAcceptors(mol),
            'NumHDonors': Descriptors.NumHDonors(mol),
            'TPSA': Descriptors.TPSA(mol)
        }
        return descriptors
    else:
        return None


# Apply the descriptor calculation function
data_cleaned['Descriptors'] = data_cleaned['combined_smiles'].apply(calculate_descriptors)

# Expand the descriptor columns
descriptor_df = pd.json_normalize(data_cleaned['Descriptors'])
data_cleaned = pd.concat([data_cleaned.drop(columns=['Descriptors']), descriptor_df], axis=1)

# Remove rows with invalid SMILES that resulted in None descriptors
data_cleaned = data_cleaned.dropna()

# Function to calculate molecular fingerprints
def calculate_fingerprints(smiles):
    mol = Chem.MolFromSmiles(smiles)
    if mol:
        fingerprints = MACCSkeys.GenMACCSKeys(mol)
        return list(fingerprints)
    else:
        return None


# Apply the fingerprint calculation function
data_cleaned['Fingerprints'] = data_cleaned['combined_smiles'].apply(calculate_fingerprints)

# Convert the fingerprints into a DataFrame
fingerprint_df = data_cleaned['Fingerprints'].apply(lambda x: pd.Series(x)).astype(int)
data_cleaned = pd.concat([data_cleaned.drop(columns=['Fingerprints']), fingerprint_df], axis=1)

# Remove rows with invalid SMILES that resulted in None fingerprints
data_cleaned = data_cleaned.dropna()

# Display the updated data with descriptors and fingerprints
print(data_cleaned.head())

                                 combined_smiles  Transfection_Efficiency  \
0  O=P(OCCCC)([O-])OCC[NH2+]CCCC.CCCCCCCCCCCCCCC...                     0.0
1  O=P(OCCCCC)([O-])OCC[NH2+]CCCC.CCCCCCCCCCCCCC...                     0.0
2  O=P(OCCCCCC)([O-])OCC[NH2+]CCCC.CCCCCCCCCCCCC...                     0.0
3  O=P(OCCCCCCC)([O-])OCC[NH2+]CCCC.CCCCCCCCCCCC...                     0.0
4  O=P(OCCCCCCCC)([O-])OCC[NH2+]CCCC.CCCCCCCCCCC...                     0.0

   Molecular_Weight  Number_of_Rings     logP  Polar_Surface_Area  NumHDonors  \
0          3685.201              4.0   31.719              575.85         2.0
1          3699.228              4.0   32.109              575.85         2.0
2          3713.255              4.0   32.499              575.85         2.0
3          3727.282              4.0   32.889              575.85         2.0
4          3741.309              4.0   33.279              575.85         2.0

   NumHAcceptors  NumRotatableBonds  unsaturation  ...  157  158  159  160  \
0           56.0              214.0           4.0  ...    1    1    1    1
1           56.0              215.0           4.0  ...    1    1    1    1
2           56.0              216.0           4.0  ...    1    1    1    1
3           56.0              217.0           4.0  ...    1    1    1    1
4           56.0              218.0           4.0  ...    1    1    1    1

   161  162  163  164  165  166
0    1    0    1    1    1    1
1    1    0    1    1    1    1
2    1    0    1    1    1    1
3    1    0    1    1    1    1
4    1    0    1    1    1    1

[5 rows x 185 columns]
```

Appendix B Model Development

- Annotated Code Snippet for the Supervised Learning Model RFC

K. W. Ramadurai, A. Banerjee, *Machine Learning-Driven Rational Design in Nanomedicine*, SpringerBriefs in Bioengineering,
https://doi.org/10.1007/978-3-032-04012-1

```python
import pandas as pd
from sklearn.ensemble import RandomForestClassifier
from sklearn.model_selection import cross_val_score
from sklearn.preprocessing import StandardScaler
import numpy as np

file_path = 'Finalized_ConcatenatedDissertation_LNP Data.csv'
data = pd.read_csv(file_path)

print(data.head())

# Data Preprocessing: Extract features and target variable
X = data.drop(columns=['combined_smiles', 'Transfection_Efficiency'])
y = data['Transfection_Efficiency']

# Standardize the features
scaler = StandardScaler()
X_scaled = scaler.fit_transform(X)

# Initialize the RandomForestClassifier
rf_classifier = RandomForestClassifier(n_estimators=100, random_state=42)

# Perform 5-fold cross-validation
cv_scores = cross_val_score(rf_classifier, X_scaled, y, cv=5)

# Output the cross-validation scores
print("Cross-validation scores:", cv_scores)
print("Mean accuracy:", np.mean(cv_scores))
print("Standard deviation of accuracy:", np.std(cv_scores))
```

```
                                 combined_smiles  Transfection_Efficiency  \
0  O=P(OCCCC)([O-])OCC[NH2+]CCCC.CCCCCCCCCCCCCCCC...                        0
1  O=P(OCCCCC)([O-])OCC[NH2+]CCCC.CCCCCCCCCCCCCCC...                        0
2  O=P(OCCCCCC)([O-])OCC[NH2+]CCCC.CCCCCCCCCCCCCC...                        0
3  O=P(OCCCCCCC)([O-])OCC[NH2+]CCCC.CCCCCCCCCCCCC...                        0
4  O=P(OCCCCCCCC)([O-])OCC[NH2+]CCCC.CCCCCCCCCCCC...                        0

   Molecular_Weight  Number_of_Rings     logP  Polar_Surface_Area  NumHDonors  \
0          3685.201                4   31.719              575.85           2
1          3699.228                4   32.109              575.85           2
2          3713.255                4   32.499              575.85           2
3          3727.282                4   32.889              575.85           2
4          3741.309                4   33.279              575.85           2

   NumHAcceptors  NumRotatableBonds  unsaturation  ionizable  helper_lipid  \
0             56                214             4          1             1
1             56                215             4          1             1
2             56                216             4          1             1
3             56                217             4          1             1
4             56                218             4          1             1

   pegylated
0          1
1          1
2          1
3          1
4          1
Cross-validation scores: [0.824       0.824       0.78225806 0.81451613 0.48387097]
Mean accuracy: 0.7457290322580644
Standard deviation of accuracy: 0.13182440564249429
```

```python
import pandas as pd
import numpy as np
from sklearn.ensemble import RandomForestClassifier
from sklearn.model_selection import train_test_split, StratifiedKFold
from sklearn.metrics import roc_curve, roc_auc_score, confusion_matrix, ConfusionMatrixDisplay
import matplotlib.pyplot as plt
from sklearn.preprocessing import StandardScaler


data_path = 'Finalized_ConcatenatedDissertation_LNP Data.csv'
data = pd.read_csv(data_path)

# Extract features and target variable
X = data.drop(columns=['Transfection_Efficiency', 'combined_smiles'])
y = data['Transfection_Efficiency']

# Standardize features
scaler = StandardScaler()
X_scaled = scaler.fit_transform(X)

# Initialize StratifiedKFold with 5 splits
skf = StratifiedKFold(n_splits=5, shuffle=True, random_state=42)

# Initialize the RFC model
rf_classifier = RandomForestClassifier(n_estimators=100, random_state=42)

# Initialize lists to store the results
test_preds = []
test_probs = []
test_true = []

# Perform 5-fold cross-validation and collect the test set predictions
for train_index, test_index in skf.split(X_scaled, y):
    X_train, X_test = X_scaled[train_index], X_scaled[test_index]
    y_train, y_test = y[train_index], y[test_index]

    # Train the model
    rf_classifier.fit(X_train, y_train)

    # Predict probabilities and labels for the test set
    y_pred_proba = rf_classifier.predict_proba(X_test)[:, 1]
    y_pred = rf_classifier.predict(X_test)

    # Store results
    test_probs.extend(y_pred_proba)
    test_preds.extend(y_pred)
    test_true.extend(y_test)

# Convert lists to numpy arrays
test_probs = np.array(test_probs)
test_preds = np.array(test_preds)
test_true = np.array(test_true)

# Calculate ROC curve and AUC
fpr, tpr, thresholds = roc_curve(test_true, test_probs)
roc_auc = roc_auc_score(test_true, test_probs)
print(f'ROC AUC Score: {roc_auc:.4f}')

# Plot ROC curve
plt.figure(figsize=(10, 6))
plt.plot(fpr, tpr, label=f'ROC curve (area = {roc_auc:.2f})')
plt.plot([0, 1], [0, 1], 'k--')
plt.xlim([0.0, 1.0])
plt.ylim([0.0, 1.05])
plt.xlabel('False Positive Rate')
plt.ylabel('True Positive Rate')
plt.title('Receiver Operating Characteristic (ROC) Curve')
plt.legend(loc="lower right")
plt.show()

# Generate and plot confusion matrix
cm = confusion_matrix(test_true, test_preds)
disp = ConfusionMatrixDisplay(confusion_matrix=cm)
disp.plot(cmap=plt.cm.Blues)
plt.title('Confusion Matrix')
plt.show()
```

```
ROC AUC Score: 0.9153
```

- Code for the SSL with Pseudo-Labeling and Model Evaluation (below on the next page)

```python
import pandas as pd
from sklearn.preprocessing import StandardScaler
from sklearn.model_selection import train_test_split, cross_val_score
from sklearn.ensemble import RandomForestClassifier
import numpy as np
from sklearn.metrics import roc_curve, roc_auc_score, confusion_matrix, ConfusionMatrixDisplay
import matplotlib.pyplot as plt
import seaborn as sns

file_path = 'Finalized_ConcatenatedDissertation_LNP Data.csv'
data = pd.read_csv(file_path)

# Preprocessing the data: Extract features and target variable
X = data.drop(columns=['combined_smiles', 'Transfection_Efficiency'])
y = data['Transfection_Efficiency']

# Standardize the features
scaler = StandardScaler()
X_scaled = scaler.fit_transform(X)

# Split the data into training and test sets
X_train, X_test, y_train, y_test = train_test_split(X_scaled, y, test_size=0.2, random_state=42, stratify=y)

# Split the training data further into labeled and unlabeled sets
X_labeled, X_unlabeled, y_labeled, _ = train_test_split(X_train, y_train, test_size=0.8, random_state=42, stratify=y_train)

# Initialize the base classifier
base_classifier = RandomForestClassifier(n_estimators=100, random_state=42)

# Train the classifier on the labeled data
base_classifier.fit(X_labeled, y_labeled)

# Parameters for iterative pseudo-labeling
max_iterations = 10
confidence_threshold = 0.9

# Iterative pseudo-labeling process
for iteration in range(max_iterations):
    # Predict probabilities for the unlabeled data
    probas = base_classifier.predict_proba(X_unlabeled)
    max_probas = np.max(probas, axis=1)
    pseudo_labels = np.argmax(probas, axis=1)

    # Select the most confident predictions
    confident_indices = np.where(max_probas >= confidence_threshold)[0]

    # If no confident predictions, break the loop
    if len(confident_indices) == 0:
        break

    # Add the confident pseudo-labeled data to the labeled dataset
    X_labeled = np.vstack((X_labeled, X_unlabeled[confident_indices]))
    y_labeled = np.hstack((y_labeled, pseudo_labels[confident_indices]))

    # Remove the confident pseudo-labeled data from the unlabeled dataset
    X_unlabeled = np.delete(X_unlabeled, confident_indices, axis=0)

    # Re-train the classifier on the expanded labeled dataset
    base_classifier.fit(X_labeled, y_labeled)

# Evaluate the final model on the test set
test_score = base_classifier.score(X_test, y_test)
print(f"Test Set Accuracy: {test_score}")

# Perform 5-fold cross-validation on the expanded labeled dataset
cv_scores_ssl = cross_val_score(base_classifier, X_labeled, y_labeled, cv=5)
print(f"5-Fold CV Scores: {cv_scores_ssl}")
print(f"Mean CV Score: {np.mean(cv_scores_ssl)}")
print(f"Standard Deviation CV Score: {np.std(cv_scores_ssl)}")

# Predict probabilities for the test set
y_test_probas = base_classifier.predict_proba(X_test)[:, 1]

# Calculate the ROC curve
fpr, tpr, thresholds = roc_curve(y_test, y_test_probas)

# Calculate the AUC score
auc_score = roc_auc_score(y_test, y_test_probas)
print(f"AUC Score: {auc_score}")

# Predict the test set labels
y_test_pred = base_classifier.predict(X_test)

# Calculate the confusion matrix
cm = confusion_matrix(y_test, y_test_pred)

# Plot the ROC curve
plt.figure(figsize=(10, 6))
plt.plot(fpr, tpr, label=f'ROC curve (area = {auc_score:.2f})')
plt.plot([0, 1], [0, 1], 'k--')
plt.xlim([0.0, 1.0])
plt.ylim([0.0, 1.05])
plt.xlabel('False Positive Rate')
plt.ylabel('True Positive Rate')
plt.title('Receiver Operating Characteristic (ROC) Curve')
plt.legend(loc="lower right")
plt.show()

# Plot confusion matrix
cm_display = ConfusionMatrixDisplay(confusion_matrix=cm)
cm_display.plot()
plt.title('Confusion Matrix')
plt.show()

# Analyze feature importance
feature_importances = base_classifier.feature_importances_
features = X.columns

# Create a DataFrame for feature importances
importance_df = pd.DataFrame({'Feature': features, 'Importance': feature_importances})

# Sort the DataFrame by importance
importance_df = importance_df.sort_values(by='Importance', ascending=False)

# Plot feature importances
plt.figure(figsize=(12, 8))
sns.barplot(x='Importance', y='Feature', data=importance_df)
plt.title('Feature Importances from the Random Forest Classifier')
plt.show()


Test Set Accuracy: 0.864
5-Fold CV Scores: [0.85897436 0.94805195 1.         1.         0.97402597]
Mean CV Score: 0.9562104562104562
Standard Deviation CV Score: 0.052295049147520624
AUC Score: 0.8968253968253969
```

- Code for SMOTE Implementation and Hyperparameter Tuning & Model Evaluation (below on next page)

```python
import pandas as pd
from sklearn.preprocessing import StandardScaler
from sklearn.model_selection import train_test_split, GridSearchCV, cross_val_score
from sklearn.ensemble import RandomForestClassifier
from sklearn.metrics import roc_curve, roc_auc_score, confusion_matrix, ConfusionMatrixDisplay
from imblearn.over_sampling import SMOTE
import numpy as np
import matplotlib.pyplot as plt
import seaborn as sns

# Load the dataset
file_path = 'Finalized_ConcatenatedDissertation_LNP Data.csv'
data = pd.read_csv(file_path)

# Preprocessing the data: Extract features and target variable
X = data.drop(columns=['combined_smiles', 'Transfection_Efficiency'])
y = data['Transfection_Efficiency']

# Standardize the features
scaler = StandardScaler()
X_scaled = scaler.fit_transform(X)

# Split the data into training and test sets
X_train, X_test, y_train, y_test = train_test_split(X_scaled, y, test_size=0.2, random_state=42, stratify=y)

# Split the training data further into labeled and unlabeled sets
X_labeled, X_unlabeled, y_labeled, _ = train_test_split(X_train, y_train, test_size=0.8, random_state=42, stratify=y_train)

# Apply SMOTE to the labeled data
smote = SMOTE(random_state=42)
X_labeled_smote, y_labeled_smote = smote.fit_resample(X_labeled, y_labeled)

# Initialize the base classifier
base_classifier = RandomForestClassifier(n_estimators=100, random_state=42)

# Train the classifier on the labeled data
base_classifier.fit(X_labeled_smote, y_labeled_smote)

# Parameters for iterative pseudo-labeling
max_iterations = 10
confidence_threshold = 0.9

# Iterative pseudo-labeling process
for iteration in range(max_iterations):
    # Predict probabilities for the unlabeled data
    probas = base_classifier.predict_proba(X_unlabeled)
    max_probas = np.max(probas, axis=1)
    pseudo_labels = np.argmax(probas, axis=1)

    # Select the most confident predictions
    confident_indices = np.where(max_probas >= confidence_threshold)[0]

    # If no confident predictions, break the loop
    if len(confident_indices) == 0:
        break

    # Add the confident pseudo-labeled data to the labeled dataset
    X_labeled_smote = np.vstack((X_labeled_smote, X_unlabeled[confident_indices]))
    y_labeled_smote = np.hstack((y_labeled_smote, pseudo_labels[confident_indices]))

    # Remove the confident pseudo-labeled data from the unlabeled dataset
    X_unlabeled = np.delete(X_unlabeled, confident_indices, axis=0)

    # Re-train the classifier on the expanded labeled dataset
    base_classifier.fit(X_labeled_smote, y_labeled_smote)

# Hyperparameter tuning using GridSearchCV
param_grid = {
    'n_estimators': [100, 200, 300],
    'max_features': ['auto', 'sqrt', 'log2'],
    'max_depth': [None, 10, 20, 30],
    'min_samples_split': [2, 5, 10],
    'min_samples_leaf': [1, 2, 4],
    'bootstrap': [True, False]
}

grid_search = GridSearchCV(estimator=base_classifier, param_grid=param_grid, cv=5, n_jobs=-1, verbose=2)
grid_search.fit(X_labeled_smote, y_labeled_smote)
best_params = grid_search.best_params_
print(f"Best parameters found: {best_params}")

# Train classifier with best parameters
best_rf_smote = RandomForestClassifier(**best_params, random_state=42)
best_rf_smote.fit(X_labeled_smote, y_labeled_smote)

# Evaluate the final model on test set
test_score_smote = best_rf_smote.score(X_test, y_test)
print(f"Test Set Accuracy with SMOTE: {test_score_smote}")

# 5-fold cross-validation on the SMOTE-balanced labeled dataset
cv_scores_ssl_best_smote = cross_val_score(best_rf_smote, X_labeled_smote, y_labeled_smote, cv=5)
print(f"5-Fold CV Scores with SMOTE: {cv_scores_ssl_best_smote}")
print(f"Mean CV Score with SMOTE: {np.mean(cv_scores_ssl_best_smote)}")
print(f"Standard Deviation CV Score with SMOTE: {np.std(cv_scores_ssl_best_smote)}")

# Predict probabilities for test set
y_test_probas_smote = best_rf_smote.predict_proba(X_test)[:, 1]

# Calculate ROC curve
fpr_smote, tpr_smote, thresholds_smote = roc_curve(y_test, y_test_probas_smote)

# Calculate AUC score
auc_score_smote = roc_auc_score(y_test, y_test_probas_smote)
print(f"AUC Score with SMOTE: {auc_score_smote}")

# Predict test set labels
y_test_pred_smote = best_rf_smote.predict(X_test)

# Calc confusion matrix
cm_smote = confusion_matrix(y_test, y_test_pred_smote)

# Plot ROC curve
plt.figure(figsize=(10, 6))
plt.plot(fpr_smote, tpr_smote, label=f'ROC curve with SMOTE (area = {auc_score_smote:.2f})')
plt.plot([0, 1], [0, 1], 'k--')
plt.xlim([0.0, 1.0])
plt.ylim([0.0, 1.05])
plt.xlabel('False Positive Rate')
plt.ylabel('True Positive Rate')
plt.title('Receiver Operating Characteristic (ROC) Curve with SMOTE')
plt.legend(loc="lower right")
plt.show()

# Plot confusion matrix
cm_display_smote = ConfusionMatrixDisplay(confusion_matrix=cm_smote)
cm_display_smote.plot()
plt.title('Confusion Matrix with SMOTE')
plt.show()
```

- ## Code for XGBOOST Model & Model Evaluation

```python
import pandas as pd
from imblearn.over_sampling import SMOTE
from sklearn.model_selection import train_test_split, GridSearchCV
from sklearn.metrics import accuracy_score, precision_score, recall_score, f1_score, roc_auc_score
import xgboost as xgb

data_path = 'Finalized_ConcatenatedDissertation_LNP Data.csv'
data = pd.read_csv(data_path)
X = data.drop(columns=['combined_smiles', 'Transfection_Efficiency'])
y = data['Transfection_Efficiency']

# Apply SMOTE
smote = SMOTE(random_state=42)
X_resampled, y_resampled = smote.fit_resample(X, y)

# Split resampled data into training and testing sets
X_train, X_test, y_train, y_test = train_test_split(X_resampled, y_resampled, test_size=0.2, random_state=42, stratify=y_resampled)

# Define XGBoost classifier
xgb_model = xgb.XGBClassifier(use_label_encoder=False, eval_metric='logloss')

# Set up the parameter grid for tuning
param_grid = {
    'n_estimators': [50, 100, 200],
    'learning_rate': [0.01, 0.1, 0.2],
    'max_depth': [3, 5, 7],
    'subsample': [0.8, 1.0],
    'colsample_bytree': [0.8, 1.0]
}

# Set up GridSearchCV
grid_search = GridSearchCV(estimator=xgb_model, param_grid=param_grid, cv=5, scoring='accuracy', n_jobs=-1, verbose=2)

# Fit model
grid_search.fit(X_train, y_train)

# Best model
best_xgb_model = grid_search.best_estimator_

# Print best parameters
print(f"Best parameters found: {grid_search.best_params_}")

# Predict on the test set
y_pred = best_xgb_model.predict(X_test)
y_pred_proba = best_xgb_model.predict_proba(X_test)[:, 1]

# Calculate evaluation metrics
accuracy = accuracy_score(y_test, y_pred)
precision = precision_score(y_test, y_pred)
recall = recall_score(y_test, y_pred)
f1 = f1_score(y_test, y_pred)
roc_auc = roc_auc_score(y_test, y_pred_proba)

print(f"Accuracy: {accuracy:.4f}")
print(f"Precision: {precision:.4f}")
print(f"Recall: {recall:.4f}")
print(f"F1 Score: {f1:.4f}")
print(f"AUC Score: {roc_auc:.4f}")
```

```
Best parameters found: {'colsample_bytree': 0.8, 'learning_rate': 0.1, 'max_depth': 3, 'n_estimators': 200, 'subsample': 1.0}
Accuracy: 0.9538
Precision: 0.9314
Recall: 0.9794
F1 Score: 0.9548
AUC Score: 0.9862
```

Appendix C

The complete code for all analyses conducted in this dissertation was developed in Anaconda and is accessible via the following Google Drive link: **mRNA Nanomedicine Analysis Code Files**.

Here are the steps to access and run the .ipynb (Jupyter Notebook) and .csv (data) files from the Google Drive link in Jupyter Notebook:

1. **Download the Files**: Click on the Google Drive link, then right-click on each file, select "Download," and download both the `.ipynb` (Jupyter Notebook) and `.csv` (data) files.
2. **Install Jupyter Notebook**: If Jupyter Notebook is not installed, first install Python, then run the "command `pip install jupyterlab`" in the terminal or command prompt. Using Anaconda is recommended, as it comes with Jupyter Notebook pre-installed.
3. **Open Jupyter Notebook**: Once Jupyter Notebook is installed, open it by running the command "`jupyter notebook`" in the terminal or command prompt. This will launch the Jupyter interface in the browser.
4. **Access the Notebooks**: In the Jupyter interface, navigate to the folder where the downloaded `.ipynb` files are stored. Click on a notebook file to open it, and execute the code cells by selecting each cell and pressing `Shift + Enter`.
5. **Load the Data Files**: The data can be loaded into the Jupyter Notebook using the Pandas library, and the notebooks are configured to read the data files.

References

Adams, D., Gonzalez-Duarte, A., O'Riordan, W. D., Yang, C. C., Ueda, M., Kristen, A. V., Tournev, I., Schmidt, H. H., Coelho, T., Berk, J. L., & Lin, K. P. (2018). Patisiran, an RNAi therapeutic, for hereditary transthyretin amyloidosis. *New England Journal of Medicine, 379*(1), 11–21.

Alcimed. (2021). *Artificial intelligence at the service of drug repositioning?*. [online] Alcimed. Available at: https://www.alcimed.com/en/insights/artificial-intelligence-at-the-service-of-drug-repositioning/ [Accessed 20 June. 2024].

Alfagih, I. M., Aldosari, B., AlQuadeib, B., Almurshedi, A., & Alfagih, M. M. (2020). Nanoparticles as adjuvants and nanodelivery systems for mRNA-based vaccines. *Pharmaceutics, 13*(1), 45.

Ali, N. (2021). *Toward data science in biophotonics: Biomedical investigations-based study (Doctoral dissertation)*.

Ali, Z. A., Abduljabbar, Z. H., Taher, H. A., Sallow, A. B., & Almufti, S. M. (2023a). Exploring the power of extreme gradient boosting algorithm in machine learning: A review. *Academic Journal of Nawroz University, 12*(2), 320–334.

Ali, A., Al-Rimy, B. A. S., Tin, T. T., Altamimi, S. N., Qasem, S. N., & Saeed, F. (2023b). Empowering precision medicine: Unlocking revolutionary insights through blockchain-enabled federated learning and electronic medical records. *Sensors, 23*(17), 7476.

Alipour, E., Halverson, D., McWhirter, S., & Walker, G. C. (2017). Phospholipid bilayers: Stability and encapsulation of nanoparticles. *Annual Review of Physical Chemistry, 68*, 261–283.

Arazo, E., Ortego, D., Albert, P., O'Connor, N. E., & McGuinness, K. (2020, July). Pseudo-labeling and confirmation bias in deep semi-supervised learning. In *2020 international joint conference on neural networks (IJCNN)* (pp. 1–8). IEEE.

Avrithis, Y. (2021). *Iterative label cleaning for semi-supervised learning*. Ph.D. dissertation, National and Kapodistrian University of Athens.

Badie-Modiri, A., & Kivelä, M. (2023). Reticula: A temporal network and hypergraph analysis software package. *SoftwareX, 21*, 101301.

Bannigan, P., Aldeghi, M., Bao, Z., Häse, F., Aspuru-Guzik, A., & Allen, C. (2021). Machine learning directed drug formulation development. *Advanced Drug Delivery Reviews, 175*, 113806.

Beck, J. D., Reidenbach, D., Salomon, N., Sahin, U., Türeci, Ö., Vormehr, M., & Kranz, L. M. (2021). mRNA therapeutics in cancer immunotherapy. *Molecular Cancer, 20*(1), 69.

Butcher, B., Huang, V. S., Robinson, C., Reffin, J., Sgaier, S. K., Charles, G., & Quadrianto, N. (2021). Causal datasheet for datasets: An evaluation guide for real-world data analysis and data collection design using Bayesian networks. *Frontiers in Artificial Intelligence, 4*, 612551.

K. W. Ramadurai, A. Banerjee, *Machine Learning-Driven Rational Design in Nanomedicine*, SpringerBriefs in Bioengineering, https://doi.org/10.1007/978-3-032-04012-1

Cánovas-Segura, B., Morales, A., Martinez-Carrasco, A. L., Campos, M., Juarez, J. M., Rodríguez, L. L., & Palacios, F. (2019, June). Improving interpretable prediction models for antimicrobial resistance. In *2019 IEEE 32nd international symposium on computer-based medical systems (CBMS)* (pp. 543–546). IEEE.

Capecchi, A., Probst, D., & Reymond, J. L. (2020). One molecular fingerprint to rule them all: Drugs, biomolecules, and the metabolome. *Journal of Cheminformatics, 12*, 1–15.

Carissimi, G., Montalbán, M. G., Fuster, M. G., & Víllora, G. (2021). Nanoparticles as drug delivery systems. *21st century nanostructured materials—Physics, chemistry, classification, and emerging applications in industry, biomedicine, and agriculture.*

Carrington, A.M., Manuel, D.G., Fieguth, P.W., Ramsay, T., Osmani, V., Wernly, B., Bennett, C., Hawken, S., McInnes, M., Magwood, O. & Sheikh, Y. (2021). Deep ROC analysis and AUC as balanced average accuracy to improve model selection, understanding and interpretation. *arXiv preprint arXiv:2103.11357.*

Carter, B. M. (2023). *Interpretations of machine learning and their application to therapeutic design* (Doctoral dissertation, Massachusetts Institute of Technology).

Chatterjee, S., Kon, E., Sharma, P., & Peer, D. (2024). Endosomal escape: A bottleneck for LNP-mediated therapeutics. *Proceedings of the National Academy of Sciences, 121*(11), e2307800120.

Chaudhary, N., Weissman, D., & Whitehead, K. A. (2021). mRNA vaccines for infectious diseases: Principles, delivery and clinical translation. *Nature Reviews Drug Discovery, 20*(11), 817–838.

Chen, T., & Guestrin, C. (2016, August). Xgboost: A scalable tree boosting system. In *Proceedings of the 22nd acm sigkdd international conference on knowledge discovery and data mining* (pp. 785–794).

Chen, S., Huang, X., Xue, Y., Álvarez-Benedicto, E., Shi, Y., Chen, W., Koo, S., Siegwart, D. J., Dong, Y., & Tao, W. (2023). Nanotechnology-based mRNA vaccines. *Nature Reviews Methods Primers, 3*(1), 63.

Cheng, X., & Lee, R. J. (2016). The role of helper lipids in lipid nanoparticles (LNPs) designed for oligonucleotide delivery. *Advanced Drug Delivery Reviews, 99*, 129–137.

Cheng, L., Zhu, Y., Ma, J., Aggarwal, A., Toh, W. H., Shin, C., Sangpachatanaruk, W., Weng, G., Kumar, R., & Mao, H. Q. (2023). Machine learning elucidates design features of plasmid DNA lipid nanoparticles for cell type-preferential transfection. *bioRxiv*, 2023–2012.

Cheung, E., Xia, Y., Caporini, M. A., & Gilmore, J. L. (2022). Tools shaping drug discovery and development. *Biophysics Reviews, 3*(3), 031301.

Cornebise, M., Narayanan, E., Xia, Y., Acosta, E., Ci, L., Koch, H., Milton, J., Sabnis, S., Salerno, T., & Benenato, K. E. (2022). Discovery of a novel amino lipid that improves lipid nanoparticle performance through specific interactions with mRNA. *Advanced Functional Materials, 32*(8), 2106727.

Dacoba, T. G., Anthiya, S., Berrecoso, G., Fernández-Mariño, I., Fernández-Varela, C., Crecente-Campo, J., Teijeiro-Osorio, D., Torres Andon, F., & Alonso, M. J. (2021). Nano-Oncologicals: A Tortoise Trail reaching new avenues. *Advanced Functional Materials, 31*(44), 2009860.

De Silva, M., & Brown, D. (2023). Multispectral Plant Disease Detection with Vision Transformer–Convolutional Neural Network Hybrid Approaches. *Sensors, 23*(20), 8531.

Deng, Z., Tian, Y., Song, J., An, G., & Yang, P. (2022). mRNA vaccines: The dawn of a new era of cancer immunotherapy. *Frontiers in Immunology, 13*, 887125.

DeVries, T., & Taylor, G. W. (2017). Dataset augmentation in feature space. *arXiv preprint arXiv:1702.05538.*

Ding, D.Y., Zhang, Y., Jia, Y. & Sun, J. (2023). Machine learning-guided lipid nanoparticle design for mRNA delivery. *arXiv preprint* arXiv:2308.01402.

Dube, L., & Verster, T. (2024). Assessing the performance of machine learning models for default prediction under missing data and class imbalance: A simulation study. *ORiON, 40*(1), 1–24.

Eloy, J. O., de Souza, M. C., Petrilli, R., Barcellos, J. P. A., Lee, R. J., & Marchetti, J. M. (2014). Liposomes as carriers of hydrophilic small molecule drugs: Strategies to enhance encapsulation and delivery. *Colloids and Surfaces B: Biointerfaces, 123*, 345–363.

Evers, M. J., Kulkarni, J. A., van der Meel, R., Cullis, P. R., Vader, P., & Schiffelers, R. M. (2018). State-of-the-art design and rapid-mixing production techniques of lipid nanoparticles for nucleic acid delivery. *Small Methods, 2*(9), 1700375.

Fang, X., Lan, H., Jin, K., Gong, D., & Qian, J. (2022). Nanovaccines for cancer prevention and immunotherapy: An update review. *Cancers, 14*(16), 3842.

Fu, Q., Zhao, X., Hu, J., Jiao, Y., Yan, Y., Pan, X., Wang, X., & Jiao, F. (2025). mRNA vaccines in the context of cancer treatment: From concept to application. *Journal of Translational Medicine, 23*(1). https://doi.org/10.1186/s12967-024-06033-6

Gareev, K., Tagaeva, R., Bobkov, D., Yudintceva, N., Goncharova, D., Combs, S. E., Ten, A., Samochernych, K., & Shevtsov, M. (2023). Passing of Nanocarriers across the Histohematic barriers: Current approaches for tumor Theranostics. *Nanomaterials, 13*(7), 1140.

Ginn, C., Khalili, H., Lever, R., & Brocchini, S. (2014). PEGylation and its impact on the design of new protein-based medicines. *Future Medicinal Chemistry, 6*(16), 1829–1846.

Girin, L., Leglaive, S., Bie, X., Diard, J., Hueber, T., & Alameda-Pineda, X. (2020). Dynamical variational autoencoders: A comprehensive review. *arXiv preprint arXiv:2008.12595*.

Gómez-Aguado, I., Rodríguez-Castejón, J., Vicente-Pascual, M., Rodríguez-Gascón, A., Solinís, M. Á., & del Pozo-Rodríguez, A. (2020). Nanomedicines to deliver mRNA: State of the art and future perspectives. *Nanomaterials, 10*(2), 364.

Gönén, M. (2006). Receiver operating characteristic (ROC) curves. *SAS Users Group International (SUGI), 31*, 210–231.

Guasp, P., Reiche, C., Sethna, Z., & Balachandran, V. P. (2024). RNA vaccines for cancer: Principles to practice. *Cancer Cell*.

Guo, P., Haque, F., Hallahan, B., Reif, R., & Li, H. (2012). Uniqueness, advantages, challenges, solutions, and perspectives in therapeutics applying RNA nanotechnology. *Nucleic Acid Therapeutics, 22*(4), 226–245.

Hairani, H., Anggrawan, A., & Priyanto, D. (2023). Improvement performance of the random forest method on unbalanced diabetes data classification using Smote-Tomek link. *JOIV: International Journal on Informatics Visualization, 7*(1), 258–264.

Hao, Y., Ji, Z., Zhou, H., Wu, D., Gu, Z., Wang, D., & Ten Dijke, P. (2023). Lipid-based nanoparticles as drug delivery systems for cancer immunotherapy. *MedComm, 4*(4), e339.

He, Q., Gao, H., Tan, D., Zhang, H., & Wang, J. Z. (2022). mRNA cancer vaccines: Advances, trends and challenges. *Acta Pharmaceutica Sinica B, 12*(7), 2969–2989.

Hinkson, I. V., Madej, B., & Stahlberg, E. A. (2020). Accelerating therapeutics for opportunities in medicine: A paradigm shift in drug discovery. *Frontiers in Pharmacology, 11*. https://doi.org/10.3389/fphar.2020.00770

Ho, W., Gao, M., Li, F., Li, Z., Zhang, X. Q., & Xu, X. (2021). Next-generation vaccines: Nanoparticle-mediated DNA and mRNA delivery. *Advanced Healthcare Materials, 10*(8), 2001812.

Hoang Thi, T. T., Pilkington, E. H., Nguyen, D. H., Lee, J. S., Park, K. D., & Truong, N. P. (2020). The importance of poly (ethylene glycol) alternatives for overcoming PEG immunogenicity in drug delivery and bioconjugation. *Polymers, 12*(2), 298.

Honkala, A., Malhotra, S. V., Kummar, S., & Junttila, M. R. (2022). Harnessing the predictive power of preclinical models for oncology drug development. *Nature Reviews Drug Discovery, 21*(2), 99–114.

Hou, X., Zaks, T., Langer, R., & Dong, Y. (2021). Lipid nanoparticles for mRNA delivery. *Nature Reviews Materials, 6*(12), 1078–1094.

Hu, Z., Ott, P. A., & Wu, C. J. (2018). Towards personalized, tumour-specific, therapeutic vaccines for cancer. *Nature Reviews Immunology, 18*(3), 168–182.

Huang, X., Kong, N., Zhang, X., Cao, Y., Langer, R., & Tao, W. (2022). The landscape of mRNA nanomedicine. *Nature Medicine, 28*(11), 2273–2287.

Huang, P., Deng, H., Wang, C., Zhou, Y., & Chen, X. (2024). Cellular trafficking of nanotechnology-mediated mRNA delivery. *Advanced Materials, 36*(13), 2307822.

Jentzsch, V., Osipenko, L., Scannell, J. W., & Hickman, J. A. (2023). Costs and causes of oncology drug attrition with the example of insulin-like growth factor-1 receptor inhibitors. *JAMA Network Open, 6*(7), e2324977–e2324977.

Jeyakumar, M. (2021). *Investigating the role of CLK3 in HPV associated cancers.* University of Kent (United Kingdom).

Jia, L., & Qian, S. B. (2021). Therapeutic mRNA engineering from head to tail. *Accounts of Chemical Research, 54*(23), 4272–4282.

Joloudari, J. H., Marefat, A., Nematollahi, M. A., Oyelere, S. S., & Hussain, S. (2023). Effective class-imbalance learning based on SMOTE and convolutional neural networks. *Applied Sciences, 13*(6), 4006.

Jordon, J., Szpruch, L., Houssiau, F., Bottarelli, M., Cherubin, G., Maple, C., Cohen, S.N. & Weller, A. (2022). Synthetic data—What, why and how? *arXiv preprint arXiv:2205.03257.*

Joy, T., Schmon, S.M., Torr, P.H., Siddharth, N. & Rainforth, T. (2020). Rethinking semi-supervised learning in VAEs. *arXiv preprint arXiv:2006.10102.*

Karadayı Ataş, P. (2024). Exploring the molecular interaction of PCOS and endometrial carcinoma through novel Hyperparameter-optimized ensemble clustering approaches. *Mathematics, 12*(2), 295.

Kim, J., & Shin, H. (2013). Breast cancer survivability prediction using labeled, unlabeled, and pseudo-labeled patient data. *Journal of the American Medical Informatics Association, 20*(4), 613–618.

Kimber, T. B., Gagnebin, M., & Volkamer, A. (2021). Maxsmi: Maximizing molecular property prediction performance with confidence estimation using SMILES augmentation and deep learning. *Artificial Intelligence in the Life Sciences, 1*, 100014.

Kiriiri, G. K., Njogu, P. M., & Mwangi, A. N. (2020). Exploring different approaches to improve the success of drug discovery and development projects: A review. *Future Journal of Pharmaceutical Sciences, 6*, 1–12.

Knipe, J. M. (2014). *Multi-responsive microencapsulated nanogels for the oral delivery of small interfering RNA* (Doctoral dissertation).

Kudithipudi, D., Aguilar-Simon, M., Babb, J., Bazhenov, M., Blackiston, D., Bongard, J., Brna, A. P., Chakravarthi Raja, S., Cheney, N., Clune, J., & Daram, A. (2022). Biological underpinnings for lifelong learning machines. *Nature Machine Intelligence, 4*(3), 196–210.

Kwon, H., Ali, Z. A., & Wong, B. M. (2022). Harnessing semi-supervised machine learning to automatically predict bioactivities of per-and polyfluoroalkyl substances (PFASs). *Environmental Science & Technology Letters, 10*(11), 1017–1022.

Li, J., Cheng, K., Wang, S., Morstatter, F., Trevino, R. P., Tang, J., & Liu, H. (2017). Feature selection: A data perspective. *ACM Computing Surveys (CSUR), 50*(6), 1–45.

Li, Y., Guo, L., & Ge, Y. (2023). Pseudo labels for unsupervised domain adaptation: A review. *Electronics, 12*(15), 3325.

Li, B., Raji, I. O., Gordon, A. G., Sun, L., Raimondo, T. M., Oladimeji, F. A., Jiang, A. Y., Varley, A., Langer, R. S., & Anderson, D. G. (2024). Accelerating ionizable lipid discovery for mRNA delivery using machine learning and combinatorial chemistry. *Nature Materials, 23*, 1–7.

Lindley, S. E., Lu, Y., & Shukla, D. (2023). The experimentalist's guide to machine learning for small molecule design. *ACS Applied Bio Materials, 7*(2), 657–684.

Liu, X., Huang, P., Yang, R., & Deng, H. (2023). mRNA cancer vaccines: Construction and boosting strategies. *ACS Nano, 17*(20), 19550–19580.

Liu, Y., Huang, Y., He, G., Guo, C., Dong, J., & Wu, L. (2024). Development of mRNA lipid nanoparticles: Targeting and therapeutic aspects. *International Journal of Molecular Sciences, 25*(18), 10166–10166. https://doi.org/10.3390/ijms251810166

Lokhande, V. S., Tasneeyapant, S., Venkatesh, A., Ravi, S. N., & Singh, V. (2020). Generating accurate pseudo-labels in semi-supervised learning and avoiding overconfident predictions via hermite polynomial activations. In *Proceedings of the IEEE/CVF conference on computer vision and pattern recognition* (pp. 11435–11443).

Maharjan, R., Kim, K. H., Lee, K., Han, H. K., & Jeong, S. H. (2024). Machine learning-driven optimization of mRNA-lipid nanoparticle vaccine quality with XGBoost/Bayesian method and ensemble model approaches. *Journal of Pharmaceutical Analysis, 14*, 100996.

Majchrowska, S., Mikołajczyk, A., Ferlin, M., Klawikowska, Z., Plantykow, M. A., Kwasigroch, A. & Majek, K. (2021). Waste detection in Pomerania: Non-profit project for detecting waste in environment. *arXiv preprint* arXiv:2105.06808.

Mendes, B. B., Conniot, J., Avital, A., Yao, D., Jiang, X., Zhou, X., Sharf-Pauker, N., Xiao, Y., Adir, O., Liang, H., & Shi, J. (2022). Nanodelivery of nucleic acids. *Nature Reviews Methods Primers, 2*(1), 24.

Mendolia, I., Contino, S., Perricone, U., Ardizzone, E., & Pirrone, R. (2020). Convolutional architectures for virtual screening. *BMC Bioinformatics, 21*(Suppl 8), 310.

Meulewaeter, S., Zhang, Y., Wadhwa, A., Fox, K., Lentacker, I., Harder, K. W., Cullis, P. R., De Smedt, S. C., Cheng, M. H., & Verbeke, R. (2023). Considerations on the design of lipid-based mRNA vaccines against cancer. *Journal of Molecular Biology, 436*, 168385.

Mey, F., Clauwaert, J., Van Huffel, K., Waegeman, W., & De Mey, M. (2021). Improving the performance of machine learning models for biotechnology: The quest for deus ex machina. *Biotechnology Advances, 53*, 107858.

Meyer, R. A., Neshat, S. Y., Green, J. J., Santos, J. L., & Tuesca, A. D. (2022). Targeting strategies for mRNA delivery. *Materials Today Advances, 14*, 100240.

Mitchell, R., & Frank, E. (2017). Accelerating the XGBoost algorithm using GPU computing. *PeerJ Computer Science, 3*, e127.

Mitra, S. P. (2011). Lipid nano-particles in medicine: Production, stability and drug delivery—A review. *Journal of Surface Science and Technology, 27*(1), 15.

Moayedpour, S., Broadbent, J., Riahi, S., Bailey, M., Vu Thu, H., Dobchev, D., Balsubramani, A., Nascimento Dos Santos, R., Kogler-Anele, L., Corrochano-Navarro, A., & Li, S. (2024). Representations of lipid nanoparticles using large language models for transfection efficiency prediction. *Bioinformatics, p.btae342*.

Moore, C. M. (2022). The challenges of health inequities and AI. *Intelligence-Based Medicine, 6*, 100067.

Moreno-Rocha, L. A., Domínguez-Ramírez, A. M., Cortés-Arroyo, A. R., Bravo, G., & López-Muñoz, F. J. (2012). Antinociceptive effects of tramadol in co-administration with metamizol after single and repeated administrations in rats. *Pharmacology Biochemistry and Behavior, 103*(1), 1–5.

Mosavi, A., Shirzadi, A., Choubin, B., Taromideh, F., Hosseini, F. S., Borji, M., Shahabi, H., Salvati, A., & Dineva, A. A. (2020). Towards an ensemble machine learning model of random subspace based functional tree classifier for snow avalanche susceptibility mapping. *IEEE Access, 8*, 145968–145983.

Najdi, S. (2018). *Feature extraction and selection in automatic sleep stage classification* (Doctoral dissertation, Universidade NOVA de Lisboa (Portugal)).

Niarman, A., & Amuharnis, A. (2023). Classification of chronic kidney disease based on health care records using machine learning with support vector machine. *Journal of Tourism Sciences, Technology and Industry, 2*(2), 20–26.

Obaido, G., Mienye, I. D., Egbelowo, O. F., Emmanuel, I. D., Ogunleye, A., Ogbuokiri, B., Mienye, P., & Aruleba, K. (2024). Supervised machine learning in drug discovery and development: Algorithms, applications, challenges, and prospects. *Machine Learning with Applications, 17*, 100576.

Oh, J. H., Choi, S. P., Zhu, J. H., Kim, S. H., Park, K. N., Youn, C. S., Oh, S. H., Kim, H. J., & Park, S. H. (2021). Differences in the gray-to-white matter ratio according to different computed tomography scanners for outcome prediction in post-cardiac arrest patients receiving target temperature management. *PLoS One, 16*(10), e0258480.

Oladipo, A. O., Lebelo, S. L., & Msagati, T. A. (2023). Nanocarrier design–function relationship: The prodigious role of properties in regulating biocompatibility for drug delivery applications. *Chemico-Biological Interactions, 377*, 110466.

Oršolić, D. (2023). *Optimizing approaches and representations for predictive modeling of molecular mechanisms of action and binding affinities of bioactive molecules* (Doctoral dissertation, University of Zagreb. Faculty of Food Technology and Biotechnology. Department of Biochemical Engineering. Section for Bioinformatics).

Ouma, Y. O., Nkwae, B., Odirile, P., Moalafhi, D. B., Anderson, G., Parida, B., & Qi, J. (2024). Land-use change prediction in dam catchment using logistic regression-CA, ANN-CA and random forest regression and implications for sustainable land–water nexus. *Sustainability, 16*(4), 1699.

Ouranidis, A., Vavilis, T., Mandala, E., Davidopoulou, C., Stamoula, E., Markopoulou, C. K., Karagianni, A., & Kachrimanis, K. (2021). mRNA therapeutic modalities design, formulation and manufacturing under pharma 4.0 principles. *Biomedicine, 10*(1), 50.

Öztürk, A. A., Gündüz, A. B., & Ozisik, O. (2018). Supervised machine learning algorithms for evaluation of solid lipid nanoparticles and particle size. *Combinatorial Chemistry & High Throughput Screening, 21*(9), 693–699.

Park, J. W., Lagniton, P. N., Liu, Y., & Xu, R. H. (2021). mRNA vaccines for COVID-19: What, why and how. *International Journal of Biological Sciences, 17*(6), 1446–1460.

Patel, N., Davis, Z., Hofmann, C., Vlasak, J., Loughney, J. W., DePhillips, P., & Mukherjee, M. (2023). Development and characterization of an in vitro cell-based assay to predict potency of mRNA–LNP-based vaccines. *Vaccines, 11*(7), 1224.

Peetla, C., Jin, S., Weimer, J., Elegbede, A., & Labhasetwar, V. (2014). Biomechanics and thermodynamics of nanoparticle interactions with plasma and endosomal membrane lipids in cellular uptake and endosomal escape. *Langmuir, 30*(25), 7522–7532.

Pillai, C. S., Panico, E. E., Zou, K. H., & Filipowska, E. (2022). Patient data privacy, protected health information, and ethics of real-world evidence. In *Real-world evidence in a patient-centric digital era* (pp. 51–72). Chapman and Hall/CRC.

Piotr, K., & Rongbing, L. (2023). *Comparative analysis of sampling methods for imbalanced classification* (Master's thesis, uis).

Polack, F. P., Thomas, S. J., Kitchin, N., Absalon, J., Gurtman, A., Lockhart, S., Perez, J. L., Pérez Marc, G., Moreira, E. D., Zerbini, C., & Bailey, R. (2020). Safety and efficacy of the BNT162b2 mRNA Covid-19 vaccine. *New England Journal of Medicine, 383*(27), 2603–2615.

Powers, D. M. (2020). Evaluation: From precision, recall and F-measure to ROC, informedness, markedness and correlation. *arXiv preprint arXiv:2010.16061.*

Pozzi, D., & Caracciolo, G. (2023). Looking back, moving forward: Lipid nanoparticles as a promising frontier in gene delivery. *ACS Pharmacology & Translational Science, 6*(11), 1561–1573.

Puccetti, M., Pariano, M., Schoubben, A., Ricci, M., & Giovagnoli, S. (2023). Engineering carrier nanoparticles with biomimetic moieties for improved intracellular targeted delivery of mRNA therapeutics and vaccines. *Journal of Pharmacy and Pharmacology, p.rgad089*, 592–605.

Quinto, B. (2020). *Next-generation machine learning with spark: Covers XGBoost, LightGBM, spark NLP, distributed deep learning with keras, and more.* Apress.

Ramachandran, S., Satapathy, S. R., & Dutta, T. (2022). Delivery strategies for mRNA vaccines. *Pharmaceutical Medicine, 36*(1), 11–20.

Ramezanpour, M. (2019). Computer simulation of biomolecular Systems of Interest in bionanotechnology. *Computer Simulation, 2019*, 03–12.

Rogers, D., & Hahn, M. (2010). Extended-connectivity fingerprints. *Journal of Chemical Information and Modeling, 50*(5), 742–754.

S'ari, M., Blade, H., Brydson, R., Cosgrove, S. D., Hondow, N., Hughes, L. P., & Brown, A. (2018). Toward developing a predictive approach to assess electron beam instability during transmission electron microscopy of drug molecules. *Molecular Pharmaceutics, 15*(11), 5114–5123.

Sahoo, K., Samal, A. K., Pramanik, J., & Pani, S. K. (2019). Exploratory data analysis using python. *International Journal of Innovative Technology and Exploring Engineering, 8*(12), 4727–4735.

Sakiyama, Y. (2009). The use of machine learning and nonlinear statistical tools for ADME prediction. *Expert Opinion on Drug Metabolism & Toxicology, 5*(2), 149–169.

Santana, K., Do Nascimento, L. D., Lima e Lima, A., Damasceno, V., Nahum, C., Braga, R. C., & Lameira, J. (2021). Applications of virtual screening in bioprospecting: Facts, shifts, and perspectives to explore the chemo-structural diversity of natural products. *Frontiers in Chemistry, 9,* 662688.

Sathyamoorthy, N., & Dhanaraju, M. D. (2016). Shielding therapeutic drug carriers from the mononuclear phagocyte system: A review. *Critical Reviews™ in Therapeutic Drug Carrier Systems, 33*(6), 489–567.

Schlich, M., Palomba, R., Costabile, G., Mizrahy, S., Pannuzzo, M., Peer, D., & Decuzzi, P. (2021). Cytosolic delivery of nucleic acids: The case of ionizable lipid nanoparticles. *Bioengineering & Translational Medicine, 6*(2), e10213.

Schoenmaker, L., Witzigmann, D., Kulkarni, J. A., Verbeke, R., Kersten, G., Jiskoot, W., & Crommelin, D. J. (2021). mRNA-lipid nanoparticle COVID-19 vaccines: Structure and stability. *International Journal of Pharmaceutics, 601,* 120586.

Seger, C. (2018). *An investigation of categorical variable encoding techniques in machine learning: Binary versus one-hot and feature hashing.*

Seo, Y., Lim, H., Park, H., Yu, J., An, J., Yoo, H. Y., & Lee, T. (2023). Recent progress of lipid nanoparticles-based lipophilic drug delivery: Focus on surface modifications. *Pharmaceutics, 15*(3), 772.

Shi, Y., Zhen, X., Zhang, Y., Li, Y., Koo, S., Saiding, Q., Kong, N., Liu, G., Chen, W., & Tao, W. (2024). Chemically modified platforms for better RNA therapeutics. *Chemical Reviews., 124,* 929–1033.

Shin, T. W. (2023). *Scalable chemical tool for analyzing brain: Electron microscope-like images with fluorescent microscope* (Doctoral dissertation, Massachusetts Institute of Technology).

Singha, S., Shao, K., Ellestad, K. K., Yang, Y., & Santamaria, P. (2018). Nanoparticles for immune stimulation against infection, cancer, and autoimmunity. *ACS Nano, 12*(11), 10621–10635.

Sork, H. (2018). *Profiling and exploiting lipid-based nanoparticles in vitro and in vivo.* Karolinska Institutet (Sweden).

Sun, D., & Lu, Z. R. (2023). Structure and function of cationic and ionizable lipids for nucleic acid delivery. *Pharmaceutical Research, 40*(1), 27–46.

Sung, H., Ferlay, J., Siegel, R. L., Laversanne, M., Soerjomataram, I., Jemal, A., & Bray, F. (2021). Global cancer statistics 2020: GLOBOCAN estimates of incidence and mortality worldwide for 36 cancers in 185 countries. *CA: a Cancer Journal for Clinicians, 71*(3), 209–249.

Swetha, K., Kotla, N. G., Tunki, L., Jayaraj, A., Bhargava, S. K., Hu, H., Bonam, S. R., & Kurapati, R. (2023). Recent advances in the lipid nanoparticle-mediated delivery of mRNA vaccines. *Vaccine, 11*(3), 658.

Tarawneh, A. S., Hassanat, A. B., Almohammadi, K., Chetverikov, D., & Bellinger, C. (2020). Smotefuna: Synthetic minority over-sampling technique based on furthest neighbour algorithm. *IEEE Access, 8,* 59069–59082.

Tay, B. Q., Wright, Q., Ladwa, R., Perry, C., Leggatt, G., Simpson, F., Wells, J. W., Panizza, B. J., Frazer, I. H., & Cruz, J. L. (2021). Evolution of cancer vaccines—Challenges, achievements, and future directions. *Vaccine, 9*(5), 535.

Tenchov, R., Bird, R., Curtze, A. E., & Zhou, Q. (2021). Lipid nanoparticles—From liposomes to mRNA vaccine delivery, a landscape of research diversity and advancement. *ACS Nano, 15*(11), 16982–17015.

Testas, A. (2023). Support vector machine classification with pandas, scikit-learn, and PySpark. In C. A. Berkeley (Ed.), *Distributed machine learning with PySpark: Migrating effortlessly from pandas and Scikit-learn* (pp. 259–280). Apress.

Thomas, O. S., & Weber, W. (2019). Overcoming physiological barriers to nanoparticle delivery—Are we there yet? *Frontiers in Bioengineering and Biotechnology, 7,* 415.

Tilstra, G., Couture-Senécal, J., Lau, Y. M. A., Manning, A. M., Wong, D. S., Janaeska, W. W., Wuraola, T. A., Pang, J., & Khan, O. F. (2023). Iterative design of ionizable lipids for intramuscular mRNA delivery. *Journal of the American Chemical Society, 145*(4), 2294–2304.

Ujwal, M. L. (2021). *Systems pharmacology–Machine learning approaches in profiling oncology drug candidates* (Doctoral dissertation, Massachusetts Institute of Technology).

Vercio, L. L., Amador, K., Bannister, J. J., Crites, S., Gutierrez, A., MacDonald, M. E., Moore, J., Mouches, P., Rajashekar, D., Schimert, S., & Subbanna, N. (2020). Supervised machine learning tools: A tutorial for clinicians. *Journal of Neural Engineering, 17*(6), 062001.

Verma, A. (2023). An introduction to transfection, transfection protocol and applications. *Cell Science from Technology Networks.* Available at: https://www. technologynetworks.com/cell-science/articles/an-introduction-to-transfection-transfection-protocol-and-applications-371627 [Accessed 12 May 2024].

Vlahos, A. E., Kang, J., Aldrete, C. A., Zhu, R., Chong, L. S., Elowitz, M. B., & Gao, X. J. (2022). Protease-controlled secretion and display of intercellular signals. *Nature Communications, 13*(1), 912.

Vora, L. K., Gholap, A. D., Jetha, K., Thakur, R. R. S., Solanki, H. K., & Chavda, V. P. (2023). Artificial intelligence in pharmaceutical technology and drug delivery design. *Pharmaceutics, 15*(7), 1916.

Wang, Y., & Yu, C. (2020). Emerging concepts of nanobiotechnology in mRNA delivery. *Angewandte Chemie International Edition, 59*(52), 23374–23385.

Wang, W., Feng, S., Ye, Z., Gao, H., Lin, J., & Ouyang, D. (2022). Prediction of lipid nanoparticles for mRNA vaccines by the machine learning algorithm. *Acta Pharmaceutica Sinica B, 12*(6), 2950–2962.

Wang, B., Pei, J., Xu, S., Liu, J., & Yu, J. (2023). Recent advances in mRNA cancer vaccines: Meeting challenges and embracing opportunities. *Frontiers in Immunology, 14*, 1246682.

Wang, J., Zhu, H., Gan, J., Liang, G., Li, L., & Zhao, Y. (2024). Engineered mRNA delivery systems for biomedical applications. *Advanced Materials, 36*(15), 2308029.

Weng, Y., Li, C., Yang, T., Hu, B., Zhang, M., Guo, S., Xiao, H., Liang, X. J., & Huang, Y. (2020). The challenge and prospect of mRNA therapeutics landscape. *Biotechnology Advances, 40*, 107534.

Wigh, D. S., Goodman, J. M., & Lapkin, A. A. (2022). A review of molecular representation in the age of machine learning. *Wiley Interdisciplinary Reviews: Computational Molecular Science, 12*(5), e1603.

Wongvorachan, T., He, S., & Bulut, O. (2023). A comparison of undersampling, oversampling, and SMOTE methods for dealing with imbalanced classification in educational data mining. *Information, 14*(1), 54.

Yamamoto, M., Kurino, T., Matsuda, R., Jones, H. S., Nakamura, Y., Kanamori, T., Tsuji, A. B., Sugyo, A., Tsuda, R., Matsumoto, Y., & Sakurai, Y. (2022). Delivery of aPD-L1 antibody to ip tumors via direct penetration by ip route: Beyond EPR effect. *Journal of Controlled Release, 352*, 328–337.

Yang, L., Gong, L., Wang, P., Zhao, X., Zhao, F., Zhang, Z., Li, Y., & Huang, W. (2022). Recent advances in lipid nanoparticles for delivery of mRNA. *Pharmaceutics, 14*(12), 2682.

Yu, S. M., Hu, D. H., & Zhang, J. J. (2015). Umbelliferone exhibits anticancer activity via the induction of apoptosis and cell cycle arrest in HepG2 hepatocellular carcinoma cells. *Molecular Medicine Reports, 12*(3), 3869–3873.

Yu, J., Zhang, L., Du, S., Chang, H., Lu, K., Zhang, Z., Yu, Y., Wang, L., & Ling, Q. (2022). Pseudo-label generation and various data augmentation for semi-supervised hyperspectral object detection. In *Proceedings of the IEEE/CVF conference on computer vision and pattern recognition* (pp. 305–312).

Zang, X., Zhao, X., Hu, H., Qiao, M., Deng, Y., & Chen, D. (2017). Nanoparticles for tumor immunotherapy. *European Journal of Pharmaceutics and Biopharmaceutics, 115*, 243–256.

Zhang, X., Wei, G., Sheng, Y., Yang, J., Ye, C., & Zhang, W. (2022). Data-based polymer-unit fingerprint (PUFp): A newly accessible expression of polymer organic semiconductors for machine learning. *arXiv preprint* arXiv:2211.01583.

Zhao, Z., Zheng, L., Chen, W., Weng, W., Song, J., & Ji, J. (2019). Delivery strategies of cancer immunotherapy: Recent advances and future perspectives. *Journal of Hematology & Oncology, 12*(1), 126.

Zhou, D. W., Wang, K., Zhang, Y. A., Ma, K., Yang, X. C., Li, Z. Y., Yu, S. S., Chen, K. Z., & Qiao, S. L. (2023). mRNA therapeutics for disease therapy: Principles, delivery, and clinical translation. *Journal of Materials Chemistry B, 11*(16), 3484–3510.

Zhou, F., Huang, L., Li, S., Yang, W., Chen, F., Cai, Z., Liu, X., Xu, W., Lehto, V. P., Lächelt, U., & Huang, R. (2024, April). From structural design to delivery: mRNA therapeutics for cancer immunotherapy. *Exploration, 4*(2), 20210146.

Zhu, L., & Mahato, R. I. (2010). Lipid and polymeric carrier-mediated nucleic acid delivery. *Expert Opinion on Drug Delivery, 7*(10), 1209–1226.

Zong, Y., Lin, Y., Wei, T., & Cheng, Q. (2023). Lipid nanoparticle (LNP) enables mRNA delivery for cancer therapy. *Advanced Materials, 35*(51), 2303261.

Zou, K. H., Salem, L. A., & Ray, A. (Eds.). (2022). *Real-world evidence in a patient-centric digital era*. CRC Press.

GPSR Compliance
The European Union's (EU) General Product Safety Regulation (GPSR) is a set
of rules that requires consumer products to be safe and our obligations to
ensure this.

If you have any concerns about our products, you can contact us on

ProductSafety@springernature.com

In case Publisher is established outside the EU, the EU authorized
representative is:

Springer Nature Customer Service Center GmbH
Europaplatz 3
69115 Heidelberg, Germany